Powerful Frequencies

NEW AFRICAN HISTORIES

SERIES EDITORS: JEAN ALLMAN, ALLEN ISAACMAN, AND DEREK R. PETERSON

David William Cohen and E. S. Atieno Odhiambo, *The Risks of Knowledge*

Belinda Bozzoli, *Theatres of Struggle and the End of Apartheid*

Gary Kynoch, *We Are Fighting the World*

Stephanie Newell, *The Forger's Tale*

Jacob A. Tropp, *Natures of Colonial Change*

Jan Bender Shetler, *Imagining Serengeti*

Cheikh Anta Babou, *Fighting the Greater Jihad*

Marc Epprecht, *Heterosexual Africa?*

Marissa J. Moorman, *Intonations*

Karen E. Flint, *Healing Traditions*

Derek R. Peterson and Giacomo Macola, editors, *Recasting the Past*

Moses E. Ochonu, *Colonial Meltdown*

Emily S. Burrill, Richard L. Roberts, and Elizabeth Thornberry, editors, *Domestic Violence and the Law in Colonial and Postcolonial Africa*

Daniel R. Magaziner, *The Law and the Prophets*

Emily Lynn Osborn, *Our New Husbands Are Here*

Robert Trent Vinson, *The Americans Are Coming!*

James R. Brennan, *Taifa*

Benjamin N. Lawrance and Richard L. Roberts, editors, *Trafficking in Slavery's Wake*

David M. Gordon, *Invisible Agents*

Allen F. Isaacman and Barbara S. Isaacman, *Dams, Displacement, and the Delusion of Development*

Stephanie Newell, *The Power to Name*

Gibril R. Cole, *The Krio of West Africa*

Matthew M. Heaton, *Black Skin, White Coats*

Meredith Terretta, *Nation of Outlaws, State of Violence*

Paolo Israel, *In Step with the Times*

Michelle R. Moyd, *Violent Intermediaries*

Abosede A. George, *Making Modern Girls*

Alicia C. Decker, *In Idi Amin's Shadow*

Rachel Jean-Baptiste, *Conjugal Rights*

Shobana Shankar, *Who Shall Enter Paradise?*

Emily S. Burrill, *States of Marriage*

Todd Cleveland, *Diamonds in the Rough*

Carina E. Ray, *Crossing the Color Line*

Sarah Van Beurden, *Authentically African*

Giacomo Macola, *The Gun in Central Africa*

Lynn Schler, *Nation on Board*

Julie MacArthur, *Cartography and the Political Imagination*

Abou B. Bamba, *African Miracle, African Mirage*

Daniel Magaziner, *The Art of Life in South Africa*

Paul Ocobock, *An Uncertain Age*

Keren Weitzberg, *We Do Not Have Borders*

Nuno Domingos, *Football and Colonialism*

Jeffrey S. Ahlman, *Living with Nkrumahism*

Bianca Murillo, *Market Encounters*

Laura Fair, *Reel Pleasures*

Thomas F. McDow, *Buying Time*

Jon Soske, *Internal Frontiers*

Elizabeth W. Giorgis, *Modernist Art in Ethiopia*

Matthew V. Bender, *Water Brings No Harm*

David Morton, *Age of Concrete*

Marissa J. Moorman, *Powerful Frequencies*

Powerful Frequencies

Radio, State Power, and the Cold War in Angola, 1931–2002

Marissa J. Moorman

OHIO UNIVERSITY PRESS ATHENS

Ohio University Press, Athens, Ohio 45701
ohioswallow.com

Printed in the United States of America
Ohio University Press books are printed on acid-free paper. ∞

28 27 26 25 24 23 22 21 20 19 5 4 3 2 1

Library of Congress Cataloging-in-Publication Data

Names: Moorman, Marissa Jean, author.
Title: Powerful frequencies : radio, state power, and the cold war in Angola, 1931-2002 / Marissa J. Moorman.
Other titles: New African histories series.
Description: Athens : Ohio University Press, 2019. | Series: New African histories | Includes bibliographical references and index.
Identifiers: LCCN 2019022595 | ISBN 9780821423691 (hardcover) | ISBN 9780821423707 (paperback) | ISBN 9780821446768 (pdf)
Subjects: LCSH: Radio in politics--Angola--History--20th century. | Colonists--Angola. | National liberation movements--Angola. | Angola--History--20th century. | Angola--Politics and government--20th century.
Classification: LCC HE8697.85.A5 M66 2019 | DDC 302.2344096730904--dc23
LC record available at https://lccn.loc.gov/2019022595

For Zola Kieza Moorman Lopes
And in memory of Gerald Bender and Susan Supriano

Contents

Illustrations

Acknowledgments

This book is the result of three inspirations: my aunt Susan Supriano's work as a radio producer of her show *Steppin' Out of Babylon*; John Mowitt's class Radio and Mass Culture; and the stories Angolans recounted to me about listening to the MPLA's *Angola Combatente* and about the central role that colonial state radio played in developing Angolan music, particularly *semba* (the subject of my first book, *Intonations*). *Powerful Frequencies* has taken a long time to research and write. In that endeavor, I have received generous support from a variety of institutions and individuals.

Funding provided by the Indiana University New Frontiers in the Arts and Humanities program covered travel and research in Angola during the summer of 2008. I received generous support through the Fulbright-Hays Faculty Research Abroad Fellowship, and the ACLS (American Council of Learned Societies) Fellowship funded travel and research in Angola and Portugal during the 2010–11 academic year. Luanda is the most expensive city in the world for expatriate residents (expensive for Angolans too!), and this funding made it possible for me to live in Luanda and for my daughter to attend school there. An Indiana University Outstanding Junior Faculty Award in 2009–10 also helped. Residential fellowships from Indiana University's College Arts and Humanities Institute in the spring of 2016 and from Indiana University's Institute for Advanced Study in the spring of 2017 allowed me to develop a scattered and incomplete collection of conference papers into a book manuscript.

My gratitude for Ohio University Press and its dynamic director, Gillian Berchowitz, is enormous. Jean Allman, Allen Isaacman, Derek Peterson, and Gillian bet on this book early on. I am proud to publish again in their excellent series.

In Angola, many people opened doors, extended contacts, offered short-term accommodations, and helped me find housing, school, and after-school care for Zola. Drumond Jaime has been a steadfast friend who has provided contacts, information, introductions, and logistical support. I would not have been able to work at Rádio Nacional de Angola (RNA) without his help and connections there and at the Ministry of Social Communication. I am grateful for his generosity. The then newly appointed director of RNA, Pedro Cabral, kindly granted me permission to work in the radio station's documentation center and recording archive. Maria Helena and Antónia Borges, the women who keep the recording archive organized, let me work in their small space for a month in 2011 and kindly copied sound material for me. Journalists Reginaldo Silva and Carlos Monteiro Ferreira shared their stories and contacts with me. Both have let me prod them for more information and clarifications. I have tremendous respect for the work they do, and I value my friendships with them.

Dr. Maria Conceição Neto is always a source of bibliographic references, contacts, a rich personal experience of independence, and friendship. She is an amazing historian and a dedicated teacher, thoughtful and generous. No one does research in Angola without her help and insights. Dr. Ana Paula Tavares, poet and historian, has offered thoughtful suggestions and affirmations in general, and particularly on chapter 1. Wanda Lara and Paulo Lara at the Associação Tchiweka have opened a rich and important archive at their father's (Lúcio Lara) home. Not only did they welcome my research in the archive, but they allowed me to conduct interviews for this book there. They are doing critical work in documenting Angola's independence struggle and diversifying the voices represented. To that end, Geração 80, a team of young media producers, has taken the work of not only Lara but of others out into the world, particularly with their documentary *Independência*.

In Portugal I worked in many archives on various trips. While this book is deeply shaped by the oral historical work I did, it is also much more steeped in archival work than my first book. Archives are bureaucracies. I have worked in archives with incredibly convoluted processes in place to access information and those in which it was relatively easy. Dalila Cabrita Mateus, who passed away in 2014, helped me navigate the Torre do Tombo's PIDE archives and facilitated paying for and conveying to me electronic copies. In the Torre do Tombo, the Military History Archive, the Amílcar Cabral Documentation Center, the Hemeroteca, and the archives of Portuguese Radio and Television (RTP), archivists offered guidance in retrieving information and delivered documents.

Numerous people made living in Luanda not only possible but wonderful, and that included various trips of long and short duration. Helena "Lena" Maria Correia Serra is my dearest friend—her huge heart, her nononsense approach to life, and her fight inspire me. Cesar Augusto Wilson de Carvalho, my first friend in Angola, is family. His mother, Madalena Afonso, has welcomed me for years into her home and fed and sheltered me. Luiza Moreira (in Luanda and Johannesburg) has been generous with accommodation, friendship, and nudging: "When are you going to finish that book?" Elizabeth Ceita Vera Cruz and Jean Martial Mbah are colleagues and confidantes. Domingos Coelho and his family are dear friends and supporters. Albano Cardoso offered support, encouragement, translations, and advice. Livia Apa, Fernando Arenas, Florencia Belvedere, Paulo Flores, Victor Gama, Ananya Jahanara Kabir, Premesh Lalu, Ondjaki, Lara Pawson, Christabelle Peters, António Tomás, and Anabela Vidinhas are dear friends and have been coconspirators in Luanda, Lisbon, London, Cape Town, Naples, Bloomington, and anywhere else our paths have crossed.

Various invitations to present my work facilitated my writing and challenged my thinking. Salwa Castelo Branco and Derek Pardue at Universidade Nova de Lisboa asked me to present as I was only beginning this project. Martin Murray and Howard Stein invited me to the University of Michigan African Studies Seminar. Miles Larmer invited me to present at the St. Antony's College at Oxford African Studies Seminar. Christopher J. Lee and Priya Lal included me in a small workshop on Africa in the 1970s. Clifton Crais offered me the opportunity to circulate a chapter at Emory University's African Studies Seminar. Carlos Fernandes hosted me for a seminar at the historic Universidade Eduardo Mondlane African Studies Center. Most recently, I circulated a chapter and received feedback at Columbia University's African Studies Seminar. Pedro Aires Oliveira and Maria José Lobo Antunes asked me to present at the Seminar on Comparative History in Lisbon; and Eugénia Rodrigues and Carlos Almeida asked me to give a paper at the African History Seminar at the Universidade de Lisboa. Akin Adesokan, Cara Caddoo, John Hanson, Eileen Julien, Michelle Moyd, and Micol Seigel read and responded to an earlier version of my introduction at the Indiana University Institute for Advanced Study when I was a fellow there. My dear friend Drew Thompson invited me to Bard to present a spin-off article and also attended my talk at Columbia. Claudia Gastrow and Aharon De Grassi offered constructive criticism on chapter 3 and Christopher J. Lee on chapter 5. Kate Schroeder and Paul Schauert helped collect material from FBIS at the beginning of the project.

My fellow Angolanist researchers are crucial interlocuters. I love that our ranks are growing, and I respect the work that all of them do. A shout-out, then, to Stefanie Alisch, Jeremy Ball, Marcelo Bittencourt, Mariana Candido, Ricardo Cardoso, Delinda Collier, Sylvia Croese, Manuel Ennes Ferreira, Roquinaldo Ferreira, Claudia Gastrow, Aharon de Grassi, Sousa

Jamba, Vasco Martins, Paulo Moreira, Ricardo Soares de Oliveira, Selma Pantoja, Margarida Paredes, Lara Pawson, Justin Pearce, Didier Péclard, Jon Schubert, Ned Sublette, Jean-Michel Tali, and Jelmer Vos. Marcelo and Aharon have often shared documents with me— Thank you! And, of course, Anne Pitcher—dear friend, sometimes coauthor, and publicity agent. Anne, Jean Allman, and Florence Bernault are invaluable friends and mentors. At IU, John Hanson and Eileen Julien have been the greatest guides and colleagues since I arrived.

My fellow Minnesota graduate school colleagues Sekibakiba Peter Lekgoathi, Alda Saíde Sauté, and Eleusio Viegas organized an incredible, pathbreaking workshop titled Comparative Liberation Struggle Radio in Southern Africa. I have learned and been inspired by our two meetings (Johannesburg in 2016 and Maputo in 2017). I cite the work of fellow workshop participants, but I want to name them here: Mhoze Chikowero, Cris Chinaka, Lloyd Haziveneyi, Robert Heinze, Tshepo Moloi, Dumisani Moyo, Sfiso Ndlovu, and Makhosazana Xaba.

Over the course of researching and writing this book, I joined the blog *Africa Is a Country* (AIAC) as a writer and an editor. It has been one of my richest and most rewarding intellectual endeavors yet. The brainchild and ongoing work of Sean Jacobs, the blog is the digital force that Sean is in person. I am grateful for the opportunity to write, edit, scout, and promote *AIAC*. A shout-out to several of the amazing folks I have encountered through the blog: Oumar Ba, Jessica Blatt, Grieve Chelwa, Dan Magaziner, Elliot Ross, Jon Soske, and Boima Tucker. Andrea Meeson, too, for her editing work at the blog and her editing advice on this book.

In Lisbon friends and fellow scholars have enlivened life outside the archive: André Soares and Noé João David, Jorge Murteira (who has had my back so many times), Valentina Pibiri and Alessandro Fantoni, and Augusto Nascimento.

In Bloomington I have benefited from terrific colleagues in History, in the Media School's Cinema and Media Studies Unit, the African Studies Program, and the Black Film Center/Archive. One could not ask for better colleague-friends than John Hanson and Michelle Moyd. John's experience, wisdom, and incredible laugh have gotten me through so much. I am very happy he is here to see this book. Michelle is an inspiring and committed scholar, writing buddy, and fellow mom. Shout-outs as well to Maria Bucur, Michael Dodson, Cara Caddoo, Lara Kriegel, Jason McGraw, Amrita Myers, and Ellen Wu. In Cinema and Media Studies, Akin Adesokan, Terri Francis, Josh Malitsky, Michael Martin, and Greg Waller have encouraged me and gotten me involved in other exciting projects. Jonathan Elmer, director of the College Arts and Humanities Institute, has too. Justin Garcia, Carl Ipsen, Eileen Julien, Jen Maher, Sarah Zanti, Sandi Latcha and Matthew Guterl (at a distance), and Micol Siegel have made Bloomington home and have all taken care of Zola at one time or another. Lara and Matthew titled this book.

Kaleb Alexsander, Lauren Bernovsky, Leigh Bush, Samy Camporez, Wanda Ewing, Heather Francis, April Hennessy, Christoph Irmscher, Kim Kanney, Stacey King, Rebecca and George Mankowski, Avalon Snell, Iracema Riveira, Laura Scheiber, and Jessica Ventimiglia have all spent days and nights with Zola when I could not. In the Chicago area, my sister Daphne Monroy and my brother-in-law Leo Monroy have also cared for Zola and had my back. Anne Brynn, Laura Edwards, and Rebecca Unger help keep me grounded and in touch with nonacademics. In Luanda, Helena Maria Correia Serra, Nelson Correia Serra, Sandra Marisa Correia Joaquina Lopes, and Graciete Yolanda Serra Pinto all cared for Zola, picking her up from school, feeding her, and dropping her at after-school care when I could not.

Writing can be lonely, but it doesn't have to be. Thanks to Laura Plummer for organizing writing groups; and thanks to the NCFFDD faculty coaching and the fabulous cowriters it has brought into my life. On the first round: Laura Talamante, Deborah Bleier, Barbara Raudonis, and our coach David Cook-Martin. And on a summer two-week writing bootcamp and now for two years to follow: Dagni Bredesen and Josh Robinson. Our daily accountability helps keep me writing. I'm grateful for your compassion, your insights, and your nudging.

I am incredibly grateful to my parents, who encourage me, give me a sunny place to visit, and have taken care of Zola when I've traveled and for a month in 2018 when I pressed hard on book revisions. Walton Muyumba has shared meals, music, films, ideas, and travel with me. He is a patient friend, champion of my work, and fellow traveler. I'm grateful for his calm, coaching, editing, and love. Zola was five when I started this book and is now fifteen. She has been enrolled in schools in Angola and Portugal, trotted along happily, and sometimes less so, to those places and to conferences and winter schools in Cape Town, Paris, and Chicago. I dedicate this book to her.

Abbreviations

AC	*Angola Combatente*/Fighting Angola (MPLA's radio program broadcast from Brazzaville, Dar, and Lusaka)
ALCORA	Aliança Contra as Rebeliões em Africa (a secret joint military alliance)
CCPRA	Comissão Coordenadora do Plano de Radiodifusão de Angola/Coordinating Commission for the Plan for Angolan Radio Broadcasting
CITA	Centro de Informação e Turismo de Angola/Angolan Center for Information and Tourism
CORANGOLA	(Steering) Committee on Radio in Angola
CTT	Correios, Telégrafos e Telefones/Post, Telegraph, and Telephone
DEPPI	Departamento de Educação Político-Ideológica, Propaganda e Informação/ Department of Political-Ideological Education, Information and Propaganda (MPLA)
DGS	Direcção Geral do Segurança/General Directorate of Security
EN	Emissora Nacional/National Broadcaster (of Portugal)
EOA	Emissora Oficial de Angola/Official Angolan Broadcaster (colonial broadcaster)
FAA	Forças Armadas Angolanas/Angolan Armed Forces (after 1993 FAPLA was renamed upon the integration of UNITA troops)
FALA	Forças Armadas de Libertação de Angola/Forces for the Armed Liberation of Angola (UNITA)

FAPLA	Forças Armadas Populares de Libertação de Angola/People's Armed Forces for the Liberation of Angola (MPLA)
FNLA	Frente Nacional para a Libertação de Angola/National Front for the Liberation of Angola
FUA	Frente de Unidade Angolana/Angolan United Front
GRAE	Governo Revolucionário de Angola no Exílio/Revolutionary Government of Angola in Exile
JMPLA	Juventude do Movimento Popular de Libertação de Angola (youth wing of the MPLA)
JURA	Juventude Revolucionária de Angola/United Revolutionary Youth of Angola (UNITA)
LIMA	Liga da Mulher Angolana/Angolan Women's League (UNITA)
MFA	Movimento das Forças Armadas/(Portuguese) Armed Forces Movement
MPLA	Movimento Popular para a Libertação de Angola/Popular Movement for the Liberation of Angola
MPLA-PT	Movimento Popular para a Libertação de Angola–Partido do Trabalho/Popular Movement for the Liberation of Angola–Workers' Party
MRPLA	Movimento Revolucionário do Povo Lutador de Angola/Revolutionary Movement of the Struggling Angolan People
OMA	Organização da Mulher Angolana/Angolan Women's Organization (MPLA)
PALOPS	Países Africanos de Língua Oficial Portuguesa (Portuguese-speaking African countries)
PIDE	Polícia Internacional de Defesa do Estado/International Police for the Defense of the State
PIDE/DGS	Polícia Internacional de Defesa do Estado/Direcção Geral do Segurança; International Police for the Defense of the State/General Directorate of Security
PRA	Plano de Radiodifusão de Angola/Plan for Angolan Radio Broadcasting
PT	Partido do Trabalho/Workers' Party

RNA	Rádio Nacional de Angola/ Angolan National Radio (broadcaster since independence)
RTP	Rádio e Televisão de Portugal/Portuguese Radio and Television
SCCIA	Serviços de Centralização e Coordenação de Informações de Angola/Angolan Services for the Centralization and Coordination of Information
SINTRAL	Sindicato do Trabalhadores de Angola Livre/Free Angola Workers' Syndicate (UNITA)
SWAPO	South West Africa People's Organization
TAAG	Transportes Aéreos de Angola (Angolan national airline)
UNITA	União Nacional para a Independência Total de Angola/ National Union for the Total Independence of Angola
UNTA	União Nacional dos Trabalhadores de Angola/National Union of Angolan Workers (MPLA)
UPA	União de Populações de Angola/Union of the Peoples of Angola (predecessor to the FNLA)
VAL	*Voz de Angola Livre*/Voice of Free Angola (FNLA's radio program broadcast from Kinshasa)
VdA	*Voz de Angola*/Voice of Angola (CORANGOLA radio program with Angolan announcers)

Chronology

1931	First radio broadcast by amateur hobbyist from Benguela, Angola
1940s–1950s	White settlers open member-based radio clubs across the territory
1953	Portuguese state begins broadcasting at the Emissora Oficial de Angola
1961	Three popular revolts spark the anticolonial war—MPLA and FNLA go into exile
1963	MPLA's *Angola Combatente* program begins broadcasting from Brazzaville
1965	FNLA's *Voz de Angola Livre* program begins broadcasting from Kinshasa
1966	UNITA formed
1974	Carnation Revolution in Portugal/transition to independence in colonies
1975	11 November—Agostinho Neto declares independence in the name of the MPLA
1977	27 May—attempted coup and state purge of the party (27 de Maio)
1991	Bicesse Accords between UNITA and MPLA lead to elections in Angola in 1992

Introduction

THE COVER image of a wire-and-bead-art radio embodies some of this book's key themes. I purchased this radio on tourist-thronged Seventh Street in the Melville neighborhood of Johannesburg. The artist, Jonah, was an immigrant who fled the authoritarianism and economic collapse in his home country of Zimbabwe. The technology is stripped down and simple. It is also a piece of art and, as such, a representation of radio. The wire, beadwork, and swath of a Coca-Cola can announce radio's energy, commercialization, and global circulation in an African frame. The radio works, mechanically and aesthetically. The wire radio is whimsical. It points at itself and outward. No part of the radio is from or about Angola. But this little radio contains a regional history of decolonization, national liberation movements, people crossing borders, and white settler colonies that turned the Cold War hot in southern Africa.

Powerful Frequencies focuses on radio in Angola from the first quarter of the twentieth century to the beginning of the twenty-first century. Like a radio tower or a wire-and-bead radio made in South Africa by a Zimbabwean immigrant and then carried across the Atlantic to sit on a shelf in Bloomington, Indiana, this history exceeds those borders of space and time. While state broadcasters have national ambitions—having to do with creating a common language, politics, identity, and enemy—the analysis of radio in this book alerts us to the sub- and supranational interests and communities that are almost always at play in radio broadcasting and listening.

This is true not just for radio in Angola but more generally. Radio's history is international. Though national broadcasters may first come to mind when we think of radio (the British Broadcasting Corporation [BBC] in the United Kingdom; Central Broadcasting Service [CBS] in the United States; or Rádio Nacional de Angola [RNA]/Angolan National Radio), radio's beginnings suggest something else. From its early days in the hands of amateur and ham broadcasters, to the military's adoption of radio in World War I, to the Cold War broadcasters like Radio Free Europe, Radio Liberty, and Radio Moscow, radio has been manifestly inter- and transnational. White settlers in Angola, for example, broadcast to connect the different areas of white settlement and to speak to other settlers in the region. National liberation movement radios beamed their programs across international borders and sometimes spoke to audiences in their immediate locale. RNA station employees traveled far and wide, on the continent and throughout the Angolan territory as well as overseas, to get training and report the news. UNITA's Vorgan radio station hailed audiences in the zones it controlled while simultaneously addressing foreign listeners as a regional conflict reverberated with international geopolitics.

War and conflict shape radio's history. Radio's relationship with propaganda emerged forcefully in World War II in the hands of Germany's Nazi propagandists. It developed to new levels in the Cold War, used by both the Soviet Union and the United States (and Western European states more generally), and by countries in the Third World that fought for decolonization, at faster and slower rates. While most African countries declared their independence by the early 1960s, southern Africa's countries fought long and arduous armed liberation struggles. Southern Africa remained a bastion of white settlers. South Africa and Rhodesia, ruled by white supremacist states, as well as the Portuguese-ruled Angola and Mozambique, which also had sizable white settler populations, used Cold War discourse to protect their interests. So, too, did national liberation movements. Radio stations and counterinsurgency projects in the region rang with "creolized" versions of these bipolar vocabularies.

Writing from the other end of the African continent as decolonization and the Cold War kicked off decades earlier, Frantz Fanon observed Algeria's National Liberation Front at work and said this about radio propaganda and listenership:

> Claiming to have heard the *Voice of Algeria* was, in a certain sense, distorting the truth, but it was above all the occasion to proclaim one's clandestine participation in the essence of the Revolution. It meant making a deliberate choice, though it was not explicit during the first months, between the enemy's congenital lie and the people's own lie, which suddenly acquired a dimension of truth.[1]

Fanon emphasizes the dialogic nature of these lies—the truth of one is revealed in its opposition to the other. He highlights agency. People make a "deliberate choice" between one lie and the other, indicating a changed consciousness, a new political alertness. Belief can make a lie true. "True lie" embraces propaganda and also captures all those strategic claims and mythologies that structure official imperial and nationalist narratives wherever they are.

Powerful Frequencies: Radio, State Power, and the Cold War in Angola, 1931–2002 uses radio to recount contemporary Angolan history. It tells the story of the messiness created by old and new states, their "true lies," and the people that run state bodies. It tells the story of the meaning broadcasters and listeners make of radio and its contents. It is a particular history. It is also one that resonates more widely. First, because it is a story implicated in Portuguese imperial history, in Third World decolonization movements, and in the southern African fight against apartheid complicated further by Cold War alliances; and second, because radio is a resilient medium that can also tell us about state power, listener agency, and the people that make states work (or not).

In this book, "radio" refers to a technology with particular properties and to the institution and infrastructure of radio that make broadcasting possible. The representation of radio in other media, whether wire and beads, film, music, or literature, is also a key element to understanding radio. Since I started this project, I have made particular notice of radio in African films, photography, and literature. Media scholars call this phenomenon "remediation."[2] While that is not my focus here, the remediations are significant because they help me talk about radio's meaning as much as its mechanics and politics. I use these cultural representations of radio to open most of the book's chapters.

An analysis of radio and the state in Angola raises issues that echo in contemporary life and politics. In the United States, despite the consolidation of media in a handful of large corporations (or because of it), small, noncommercial radio is growing. A January 2018 article in the *New York Times* reported that in the northwestern United States, community radios, once the bastion of rural broadcasters, are now popping up in urban neighborhoods. Low-powered, often with small transmitters and limited broadcast range, they are a quirky, local alternative to corporate-dominated radio.[3] These community radios may embody what Bertolt Brecht imagined radio could be in the 1930s:

> Here is a proposal to give radio a new function: Radio should be converted from a distribution system to a communication system. Radio could be the most wonderful public communication system

> imaginable, a gigantic system of channels—could be, that is, if it were capable not only of transmitting but of receiving, of making the listener not only hear but also speak, not of isolating him but of connecting him. This means that radio would have to give up being a purveyor and organise the listener as purveyor.[4]

Corporate, state-run, and public service radios are predominantly distribution systems, despite talk radio's popularity and the growth of podcasting. The low-power-radio operators across the United States and many community stations on the African continent show us, though, that Brecht's proposal is neither outlandish nor dated.

Old technologies such as radio are finding new uses, not obsolescence. The interface of old and new communications technologies with humans and institutions continues to bedevil human society (think: bots). As I was writing this introduction, Mark Zuckerberg, Facebook CEO, was attempting to get ahead of a problem. The problem was the infiltration of foreign (likely Russian state-financed) hackers. In the US 2016 presidential election and again in the summer of 2018 as the country was in the heat of campaigning for midterm elections in November, hackers opened false accounts on Facebook. They mobilized people on different sides of debates over gun control, immigration, and policing. They promoted white supremacist marches and counterprotests. As a result, Americans have been debating the power of media to convince, rally, and deceive audiences. And the role of corporations, the state, and individuals in producing, regulating, and interpreting media.

Commenting on Facebook's situation, *Recode* e-magazine editor Kara Swisher said, "Their problem is that they [Facebook executives] are not trained in the humanities."[5] How and why people use technologies, and what meanings they produce and how, are not questions with technical answers. They are human problems. I couldn't agree more with Swisher. State regulation, communications policy, and even foreign infiltration all have a human dimension. This book asks questions and tells stories about both the human and the technological stakes of radio broadcasting to understand radio's power, its frequency, and why states, guerrilla movements, scattered communities, and individual listeners have turned to radio.

RADIO'S TRUE LIES

"Lie" is shorthand for "propaganda." Since the advent of radio in World War I and its fatal uses by Hitler in World War II, propaganda is how we talk about the trafficked and dangerous intersection of human behavior, mass media forms, and state power.[6] Propaganda, the dissemination of ideas by the interested and powerful, has been around much longer than that, though. Papal bulls and encyclicals are just a couple of examples of centuries-old propaganda.[7] "Lie" as a substitute for "propaganda" carries the mid-twentieth-century worries and hopes about new mass media technologies.[8] At the height of the Cold War, analysts referred to totalitarian state propaganda as "brainwashing."[9] These concerns ripple through the work of the Frankfurt School and also through the writing of proponents of public diplomacy.[10]

Approaching Angola's contemporary history through radio draws our attention to colonial and postcolonial "lies," their dissemination and effects. Throughout the book, the term "true lies" also refers to the cliché that one person's or one state's lie may be another's truth. "True lie" indicates that state propagandists and broadcasters believe what they are saying and/or the big ideas they are defending. It underscores how interpretation and narrative matter in human and political lives.

In using a specific technology to think about independence struggles and postcolonial state-making, this book reveals how communications matter to colonial and postcolonial state policies, histories, and propaganda. Focusing on radio and the effects of sound, I move between state and society actors. The everyday work of state functionaries, national movement militants, and citizens sometimes takes center stage so that we can hear the interaction at work even in top-heavy, authoritarian states, be they colonial or postcolonial.

The story I tell here transforms the standard nationalist and Cold War chronologies and spaces that structure Angolan history: the attempted coup in 1977, rather than independence in 1975; and Cuban and South African troop withdrawal from Namibia in 1989 (after the 1988 battle of Cuito Cuanavale), instead of the fall of the Berlin Wall, prove key. With radio at the center, I torque nation and sharpen region to contour a flat story of superpower- funded civil war and show that quotidian listening practices and the decisions of radio journalists shape national broadcasting as much as geopolitics. Former Angolan National Radio (RNA) employees remember, for example, creating an efficient, socialist-minded, state institution that embodied their commitments to radio professionalism, the new nation, and workplace social services even as war imposed material and sonic limits.

A BRIEF HISTORY OF THE ANGOLAN INDEPENDENCE STRUGGLE

Angola's long history contains many of the processes and themes that define current US African history syllabi: Bantu migrations, politically complex centralized states, participation and subjugation in the transatlantic slave trade, and uneven, shifting colonization that used white settlement, concessionary companies, and forced labor regimes to control African lives and extract resources. Angola's history and that of the southern African region depart from the continent's history. When other countries on the continent were busy fighting for and negotiating independence from their colonizers, the Portuguese state acted forcefully to suppress nationalists in its African colonies, and the apartheid state in South Africa had cemented extreme racial segregation in law and violently attacked resistance to it.

Portugal, even in the post-Bandung world, still needed the colonial territories to fuel its economic growth and political imagination. A newly expanded United Nations and a new zeitgeist drew attention to the Portuguese colonies. To mitigate the pressure, Prime Minister António Oliveira Salazar turned to Brazilian anthropologist Gilberto Freyre and his theory of lusotropicalism. Initially invented to explain Brazil's national cultural specificity, lusotropicalism argued that the Portuguese were naturally suited to creating harmonious, multiracial societies in the tropics. When Salazar sent Freyre on a tour of Portugal's African colonies, his experience showed otherwise. Nonetheless, Salazar turned Freyre's work into a retroactive ideology of Portugal's empire.[11]

Portuguese regime defenders insisted that Portuguese rule was an exception. Nationalists, scholars, and journalists who researched and wrote in the late 1960s and early 1970s knew better. National liberation movement intellectuals, including Amílcar Cabral, Eduardo Mondlane, and Mário Pinto de Andrade, argued that rather than a tame lusotropicalist rule, Portuguese colonialism was especially violent, anchored in forced labor regimes, in de facto segregation, and only marginally less virulent than South Africa's apartheid rule. Producing empirically grounded social scientific work, Gerald Bender, Allen Isaacman, John Marcum, John Saul, and journalists such as Augusta Conchiglia and Basil Davidson followed their lead. Theirs was also an exceptionalist argument, one meant to overturn the "colonizer's congenital lie," if you will. Differences of language and the slow nature of change in the academy preserved this exceptionalism and isolationism in the scholarship on Portuguese-speaking Africa until recently. Scholars from and of Angola, Mozambique, Cape Verde, Guinea-Bissau, and São Tome have now begun to think of the histories of these places across a range of languages and in relation to other colonial histories.[12]

Angola's late colonial period and struggle for independence, especially in the context of this book, have similarities not only with histories of other places on the African continent and especially the white settler states

(Algeria, Kenya, South Africa, and Southern Rhodesia), but also with white settler societies in other places and times, as well as with other liberation movements around the globe. Guerrilla warfare became the predominant mode of armed conflict after World War II, often as a way to demand decolonization and then to contest power in new states. Many liberation movements turned to radio to supplement their ground wars.[13]

On the African continent, 1960 was the year of independence (Ghana and Guinea preceded by gaining their independence in 1957 and 1958, respectively). Seventeen countries won their freedom from colonial overrule that year, after periods of negotiation and ongoing struggle in some places (like Cameroon) but relatively rapidly in others (like the Republic of the Congo). As Angola's neighbor, the sounds and sentiments of Congo's independence rippled across their shared border. In 1961 three revolts shook the Angolan colony and initiated the anticolonial struggle's armed phase. The year began with workers in the cotton-producing region of Malanje, in the area known as the Baixa do Kassanje, revolting against forced cotton production and protesting colonial administration of their lives and land. The Portuguese reacted with violence. On 4 February a group of people armed with machetes attacked the main prison in Luanda. Again, the state and the white population reacted violently. On 15 March, the Union of the Peoples of Angola (UPA; the predecessor of the FNLA, National Front for the Liberation of Angola) and local activists violently attacked village border posts, white-owned coffee plantations throughout northwest Angola, and commercial establishments. Portugal sent troops and reinforced police activity, spreading terror. Angolan nationalist movements went into exile and undertook armed struggle. The year 1961 put Portuguese colonial rule to the test. Not only did Angola explode, but Indian nationalists took over Goa, and anticolonial protests and wars erupted across Portugal's African colonies.

In 1961 in Angola, the Portuguese army fought the National Front for the Liberation of Angola (FNLA), headed by Holden Roberto, and the Popular Movement for the Liberation of Angola (MPLA), led by Agostinho Neto. In 1966, Jonas Savimbi broke from the FNLA and founded the National Union for the Total Independence of Angola (UNITA), adding another set of combatants and ideas. Across the fourteen years that separated the explosion of war and Angolan independence in 1975, these movements fought the Portuguese military and they fought one another. The Portuguese Armed Forces Movement (MFA) overthrew the Salazar/Caetano regime in a coup on 25 April 1974, ushering in a transition of power and intiating decolonization. The Alvor Accords, an agreement for tripartite rule and then elections in Angola, broke down. In its wake, fighting among the movements (now formalized as political parties), polarization, and the intervention of foreign powers unfurled. With the help of Cuban troops and Russian armaments, the MPLA managed to control Luanda and fight off South African troops that invaded from the south to help UNITA secure the capital. Independence emerged in a Cold War–amplified civil war. The civil war continued after Cold War actors officially exited, ending only after the Angolan Armed Forces (FAA) caught and killed Jonas Savimbi in 2002.

I explore this whole period in more detail across the book's chapters. I discuss the post-2002 developments in the epilogue. The first half of the book, chapters 1 through 3, covers the late colonial period, and the second half covers the period after independence. The themes of the first three chapters—whiteness, nervousness, and technopolitics—elaborate radio in settler life, guerrilla warfare, and counterinsurgency. The themes of chapters 4, 5, and 6—state/party consolidation, propaganda, and institution-building—expose the contested terrain of radio after independence and its use in local, regional, and Cold War battles.

The book's cadence is uneven. It might feel a bit like shortwave broadcasting's skip propagation. It tunes in and out of spaces in the colonial and independent territory, sometimes exploring relations between people and sometimes worrying about what the colonial and the postcolonial states are doing. I dash over certain periods quickly, and at other times I hover over a few months (e.g., the time period around 27 May). I am interested in the choppy, dissonant story of how Angola sounded out and what meanings people made of that, why the colonial state and the independent one invested in radio, and how liberation movements and rebel movements used it to destabilize the sonic sureties of their political foes, both near and far.

The rest of this introduction situates the book in contemporary Angolan politics, delineates sources, and explores methodological issues.

THE WHY: WHY THIS BOOK NOW?

This book originates in the memories of Angolans who tuned in to *Angola Combatente*, the radio program of the MPLA guerrilla movement. Memories of listening secretly, thick with fear, brimming with the energy of the illicit, still enliven and punctuate conversation fifty years later. As scholars and students of radio have pointed out for years, reception is a key element in conceptualizing and analyzing how radio works. The very fact of the interest in listening shown by the PIDE (International Police for the Defense of the State) recenters the state in the history of Angolan radio. Two other dynamics keep the state at the center of the book: (1) political events in Angola since the end of the civil war in 2002; and (2) the research process.

The first dynamic, the centralization of power, increasingly executivized—focused in the office of the president of the republic—demands renewed attention to Angola's postcolonial state. Scholars need to think more carefully and critically about the history of the state, its fictions, and its fantasies. Work on Angola's postwar oil boom has carefully analyzed the dynamics of postcolonial statecraft.[14] The ways in which the state and former president dos Santos have handled not just the formal political opposition since 2002, but the informal movements that have taken to the streets and to social media since March 2011, require new, more historically attentive modes of analysis. Current president João Lourenço has introduced important and popular changes made possible by concentrated executive power, but he has not yet acted to decentralize power. Studies that can analyze relations between the colonial state and its postcolonial successor; that consider how people make up and contest the state from within its institutions; and that attend to the history of media and mediation can illuminate the exercise of power. What continuities exist in institutional practices? Where did change take effective hold? What is the role of state functionaries? How and why have institutions (ministries, parties, political cultures, communications practices, etc.) shifted or not shifted? What is the relationship between capitalist practices ("the market"), the state, and the media?

The second dynamic—the research process—requires more elaboration. It exposes some of the concrete operations of how the first dynamic operates in everyday life (for example, in the life of one researcher intersecting with different state bureaucracies and functionaries). No matter how well-planned, research is unpredictable. Though seemingly fixed in documents, archives, and memories, the past echoes differently depending on where and when we encounter it. You may not get to hear it at all. Sometimes the call is faint, distortion interferes, voices slip away, translation fails. At other times, sounds boom, actions clatter, and the din of grand gestures deafens the plodding, everyday work of people.

Oral historical work is fundamental to historical research even as it has its critics.[15] In investigating the history of RNA, the national broadcaster after independence, I rely on the accounts of individuals who worked at the national station, and I often use documents from their personal archives. RNA does not maintain a documentary archive. It has a documentation center, a pre-internet relic that served as a resource library for journalists. A thin, one-notebook collection of news clippings about RNA and a copy of an important dissertation on broadcasting in Angola by historian Júlio

Mendes Lopes constitute the station's written history. The sound archive (which labels their own recordings as "RNA Historical Archive"), from which I collected hundreds of recordings, holds an incomplete collection of previously broadcast material and a scattering of files related to programming, primarily from the 1980s. The collection of recorded materials concentrates on the speeches of Angola's first and second presidents: Agostinho Neto (1975–79) and José Eduardo dos Santos (1979–2017). The RNA is now recording and storing the speeches of President João Lourenço, elected in August 2017. This official sound archive guards the most significant official pronouncements since independence. Among them are a number of MPLA Party Congress openings and closings. A few recordings from the era prior to independence exist: one MPLA *Angola Combatente* guerrilla radio broadcast, the coverage of Agostinho Neto's arrival in Luanda in February 1975, the announcement of the cease-fire between the Portuguese troops and the MPLA in 1974, speeches by Portuguese general Spínola and Agostinho Neto at the Alvor Accords, a recording of Che Guevara from January 1965 during his visit to Brazzaville, and a couple of recordings of Nito Alves before his expulsion from the MPLA.[16]

RNA computerized and digitized daily paperwork beginning in the 1990s, making much material inaccessible to me. RNA bureaucrats refused my requests to see information about personnel, simple intake documents, or statistics on staff, for example, on the basis of violating the right to privacy. Doing research for my dissertation in the late 1990s, I had regularly visited RNA to meet with broadcasters and journalists who worked with musicians and to consult the station's music collection. Entering a state radio station is never easy. But access to RNA has tightened over time. When I returned to do formal research on radio in 2010–2011, I was told I needed permission from the Ministry of Social Communication. The ministry granted my request three months after I submitted a letter describing my work. This had little to do with me or the nature of my research and everything to do with internal politics (or so my journalist friends assured me): a newly appointed director of the RNA (recently renamed PCA, president of the administrative council) and tensions between the radio station and the Ministry of Social Communication.

The PCA's office sent me to work in the documentation center and the sound archive. On the floor of the sound archive, I found piles of binders full of programming plans, some correspondence between the MPLA party headquarters and the RNA management, and many binders of radio department paperwork from the 1980s. Precisely the kind of material now stowed in computer files, saved to external hard drives, or entirely lost to posterity, it is rich in detail about the radio station's operations. Abandoned in untidy stacks, this informal archive languished, sometimes teetering next to dust-covered, exposed vinyl records and mouse-infested boxes, alongside the original instruction manuals for Phillips and Nagra equipment purchased in the 1970s and pamphlets about agricultural outreach from the late colonial period. It offered the material, gritty pleasures of the analogue in jumbled chronologies. I worked for a couple of weeks, relatively undisturbed, left to myself among the dust, bug detritus, and organograms, in the capable hands of Maria Helena and Antónia, the women who ran the recording center.

One day, the director of the documentation center came to check up on me in the makeshift photo area I had set up under a channel of natural light on a couple of old chairs. He closed the top of the binder and read the title card aloud: "Correspondence: MPLA Headquarters and RNA." "No," he said. "You cannot read or photograph this. Those documents belong to the PCA's office." He looked nervous. Extremely nervous. Two weeks of attempting to convince the PCA's associate director that I should be allowed to read the rest of the notebook ensued. Two weeks to get an appointment. Ten minutes of discussion. An ambivalent answer, with a clear meaning: no.

I presented the following argument to both the director of the documentation center and the associate director: These documents were more than twenty-five years old, the standard time before declassification in official state archives. Neither cared about archival practice. The rules and practices of this system did not overlap with the protocol of other professions or disciplines.[17] The associate director stressed the possibility of finding "secret" information. But even as I noted that this was a different government (the third republic, not the second), that the war was over, and that since I had read half the material in the binder I could promise him there was nothing juicier (and nothing less surprising) than journalists playing hooky from ideological education classes at party headquarters, he was not convinced. The past we do not know is safer. Do not meddle.

This episode underscores the danger that some Angolans believe the recent past poses to the present. It likewise points to the entangled postcolonial history of the radio and the MPLA. In no sense is RNA an independent, public broadcaster. In law, RNA exists as a public institution, but in practice, it operates as an organ of the party, albeit a complex one, shot through with contradictions and exceptions. Every level of the radio's organization has its party informers, even today. They keep tabs on the station's employees and an eye on the leadership. The radio also employs collaborators from other parties. One or two of them have been integrated into the highest echelons and serve or served on its administrative council. But organograms and bylaws aside, RNA functions according to MPLA party logic and practice. Some employees who are or were UNITA members recount being shunned and called names, even as RNA leadership invested them with significant responsibilities and enticed them with good pay.

The political tensions that structure work at RNA help explain why the director did not want files opened, especially by a foreigner. The PCA's papers, perilously stacked, left for lost in the sound archives in the RNA basement, along with the documentation center director's "no," scream a nervousness that ripples across the colonial divide. It echoes the nervousness of the PIDE and military when they listened to and transcribed guerrilla movement broadcasts. It replays the nervousness Vorgan, UNITA's radio station, produced.

THE HOW: SOURCES AND METHODOLOGY

The two single largest collections from which I draw are that of RNA in Luanda and that of the PIDE housed at the Torre do Tombo, the National Archive in Lisbon. At the Torre do Tombo, I consulted the PIDE and António Salazar collections, which hold a wealth of material on the radio, as well as on broadcasting and listening in the late colonial period. I also studied documents from the Arquivo Histórico Militar (Historical Archive of the Military); the Biblioteca Nacional (Portuguese National Library); the Centro de Informação e Desenvolvimento Amilcar Cabral (CIDAC; Amílcar Cabral Center for Information and Development); the Hemeroteca (an archive dedicated to newspapers); the Ministério de Negócios Estrangeiros (Ministry of Foreign Relations); and Rádio e Televisão Portuguesa (RTP; Portuguese Radio and Television), all located in Lisbon and storing material on the late colonial period. UNITA holds no archives since the Angolan Armed Forces (FAA) confiscated much of what they did have. Oral historical work was critical. I spoke with a number of UNITA members who had been active in radio. The Foreign Broadcast Information Service (FBIS), a division of the United States Central Intelligence Agency committed to monitoring radio and television broadcasts and newspaper reports by foreign governments and entities, offered key, if partial, sources on UNITA broadcasts.[18] Research on this project began in 2005, but I undertook the most concentrated work from 2010 to 2011 in Luanda and then again on short research trips to Luanda in 2012, 2013, and 2015, and to Lisbon in 2011, 2013, 2015, 2016, 2017, and 2018.

If oral sources such as interviews suffer the pockmarks of selective memory and reconstruction in the service of the present self and interests, secret police documents pose their own distinct set of challenges. Produced by a bureaucratic system in the throes of archive fever, they proliferate. Each memo written by a police officer is reproduced in sextuplet. Words and information papered over the void of the Salazar Estado Novo's (New State) lack of knowledge.

The government ministry born from the coup that overthrew the Salazar/Caetano dictatorships preserved the documents of the PIDE. The military junta dissolved the PIDE/DGS (International Police for the Defense of the State/General Directorate of Security) on the day of the coup, 25 April 1974. The democratic Portuguese state moved and stored PIDE documents for twenty years (some destroyed by former officers; some ruined in the process) before opening them for consultation on 26 April 1996 at the National Historical Archive. This is an archive that promotes transparency about the Salazar/Caetano Estado Novo regime but that also induces nervousness. The history of *bufos* (informers), what the colonial state knew when about which national liberation movement, about its collaborations, and the interrogations of those arrested are often politically contentious and personally tender topics. A different kind of nervousness permeates the documents about the guerrilla radios of the MPLA and FNLA. For the book's chapters on nervousness to make sense, I detail how the colonial state collected and used information.

INFORMATION COLLECTION IN THE ESTADO NOVO

Understanding the structures of information collection and distribution—and sketching the relations between the military and the secret police (PIDE), in particular during the war (1961–74)—contextualizes the surveillance files at the center of chapter 2 and the debates around the radio and counterinsurgency in chapter 3. Conflicts traversed the relations between different state bodies conducting the war. Disagreements over approach, personalities, and egos shaped the working environment.[19] The Portuguese military and intelligence bodies, like much of Portuguese colonial rule after World War II, were not exceptional in this regard. Interinstitutional relations were tense. Salazar's autocratic impulses threatened command structures. He regularly overruled legislation and plans concocted by military intelligence experts.[20]

Four government bodies collected information on the war in Angola: the Angolan Armed Forces (FAA); the Angolan Services for the Centralization and Coordination of Information (SCCIA); the Ministry of Foreign Relations (via the Department of Political Affairs, GNP—diplomatic envoys, missions); and the PIDE/DGS.[21] Miguel Jerónimo Bandeira and António Costa Pinto argue that the key institutional and legal innovations of the post–World War II period aimed to align Portuguese colonial and imperial rule with international trends. New policies and institutions emphasized developmental colonialism, first implemented in the interwar period, and adopted specialized knowledges that circulated internationally to justify and execute those programs.[22] Pressures to decolonize from within the Portuguese territories and from external, international bodies, Portugal's commitment to NATO, and the war compelled these changes.

In 1959 the Portuguese state opened a new department specifically tasked with information collection: the GNP in the Ministry of Foreign Relations. This ministry "was a keystone of the new institutional network of information and intelligence gathering, which aimed to promote a new information empire" that could respond to a postwar world in transformation.[23] The GNP stood at the apex of information. It coordinated government ministries meant to transform colonial governance into a rational, technical, modern, and scientifically based practice.[24] These government bodies eventually included the PIDE, the Ministry of Foreign Affairs; SCCIA; and the Angolan Center for Information and Tourism (CITA). Under this arrangement, information funneled in and expertise flowed out, or that was the idea. Counterinsurgency militias, built on this novel information regime, "became the backbone of the 'repressive version of the developmentalist colonial state.'"[25] Counterinsurgency depended on information and communication.

The PIDE/DGS, formed in 1954, began operating in the African territories three years later; and in 1961 the state united the PIDE/DGS in Portugal and those in the "overseas territories" into one body.[26] In 1968 the state created a new position: subdirector of the PIDE for operations in Angola and Mozambique, occupied by the infamous Aníbal São José Lopes.[27] In 1969, when Marcelo Caetano assumed power after Salazar's disabling stroke, he introduced a cosmetic change, renaming the PIDE the General Directorate of Security (DGS).[28] At the time of the Portuguese military coup in April 1974, Angola had 1,119 people working for the PIDE/DGS, more than any of the other territories.[29] Unlike in Portugal, where the Armed Forces Movement immediately dismantled the PIDE/DGS, in the colonies, it remained as the Military Information Police (Polícia de Informação Militar; PIM) through the transition period and until 1975.[30]

The Overseas Ministry and the Minister of Defense drafted the legislation creating SCCI in 1961 with offices to open in Angola, Mozambique, and Guinea.[31] The Angolan office (SCCIA) opened in 1962 with a mandate to collect, coordinate, and disseminate information relevant to policy, administration, and defense.[32] Based on training in England at the British military's intelligence school on information and counterinformation, SCCIA would centralize and circulate information.[33] SCCIA collected information from the different police forces (PIDE among them), from the Angolan Provincial Organization of Volunteers and Civil Defense (Organização Provincial de Voluntários e Defesa Civil; OPVDC), administrative authorities, military commanders, and the Overseas Ministry.[34]

The military's Second Division undertook information collection and distribution, both at the battalion level and at the higher level of the whole military region.[35] This division, which produced information on neighboring countries and nationalist movement activities, centered on strategic and operational questions. It published and circulated a PERINTREP (periodic information report) every two weeks, analyzing the information.[36] The reports included maps, topographical sketches, and aerial photos. Psychological action formed part of the Second Division's remit until 1966 when that charge passed to the planning and studies office in the military headquarters for the region.[37]

PIDE/DGS in Angola, Mozambique, and Guinea-Bissau, in distinction to its operations in the metropole, worked with the Portuguese military, even if not always well. In the metropole PIDE agents worked to hide their identity, but in the colonies, everyone identified them as civil servants and, like soldiers, as representatives of the state: they were "a pillar of the colonial system and the information they provided for the military, gained through various violent means, constituted a valuable instrument in the war."[38] Despite the PIDE/DGS's successes in supporting military operations, its capacity for information collection fell short of the challenge at hand.[39] Possessing the powers of a judicial police unit, the PIDE/DGS used arrest, confiscation, interception, and interrogation to repress the liberation movements and obtain information. In addition, the organization conducted operations with the *flechas* (African militias recruited in Angola) to undermine the movements; infiltrated, sabotaged, and assassinated movement members based in neighboring countries; and ran a network of prisons where they held and tortured political detainees, who often died in detention.[40] Using the flechas yielded some results, though it moved the PIDE out of strictly information collection and into policing operations.[41]

Recent work identifies the tensions between the various bodies of information collection central to fighting the war. SCCIA never possessed the mechanisms to coordinate information, for example, though the service did what it could, as the hub of information from the PIDE/DGS, the military, and other sources of intelligence, to share the information it received.[42] It was a part of the military but fell under the colonial governor's command. In 1962, the position of governor-general was split into two: governor-general and head of the Angolan command, creating ambiguity in SCCIA that remained unresolved throughout the war.[43] Furthermore, opposition from, even obstruction by, the PIDE limited SCCIA's effectiveness.[44] The PIDE/DGS was known to initiate operations with the flechas without informing anyone, and when rivalries developed between bodies they rarely resolved easily.[45]

The PIDE/DGS worked closely with Rhodesian and South African police forces and secret services. Relations with the Rhodesian Central Intelligence Organisation (CIO) and its director, Ken Flower, had their roots in this cooperation and information-sharing.[46] The PIDE/DGS and the South African Police (SAP) signed a protocol for information exchange and the capture of insurgents in 1955, shortly after the PIDE began work in Angola and Mozambique.[47] Relations tightened with collaboration between the PIDE/DGS and the South African Bureau of State Security (BOSS) from 1962. Daily contact via equipment offered by the SAP facilitated information exchange on the actions of the MPLA and the South West Africa People's Organization (SWAPO).[48] The Portuguese Overseas Ministry kept this alliance a secret, knowing that public knowledge of joint activities would attract criticism in a decolonizing world.[49]

Rhodesia, South Africa, and Portugal found common cause in opposing what they described as a communist threat, though the ideologies of the national liberation movements and other activist groups in the region were diverse. These three countries met regularly between 1962 and 1968 to share information. Collaboration with South Africa intensified in 1968 in eastern and southern Angola. Between 1970 and 1974, the three countries engaged in a secret, joint military alliance—ALCORA—formalizing earlier cooperation in order to defeat the liberation movements.[50] Military and police information collection and sharing, in and outside the Angolan territory, proved critical to executing the war. Archives containing the documents offer a complex and important set of resources for this period in the region.

RANGE AND RECEPTION

Throughout the book I touch on radio broadcasting range and radio reception. Settlers, the colonial state, liberation movements, the postcolonial state, and UNITA all broadcast but did not necessarily keep records about broadcasting equipment or range, nor did they collect statistics. Outside of listening to the movement radio stations, which the PIDE and military followed, most information on listenership is anecdotal. Poor and partial in coverage, though rich at times in irony, nervousness, and bravery, together these stories and written sources present a picture with some clear lines, many yawning empty spaces, and splotchy connections.

Radio broadcast range and radio reception, despite their bundling with modernity and their supposed technological simplicity, are not straightforward. Broadcast range depends not just on the strength of a transmitter but on the height of the transmission tower and on atmospheric and topographical conditions. Shortwave broadcasting and range are particularly hard to measure. Shortwave's skipping propagation, the way it drops in and out, pinging off the ionosphere, is what allows it to cover great distances, though not necessarily to blanket them in sound. FM (frequency modulation) uses "line of sight" propagation. The higher the tower, the greater the range, though transmitter strength and frequency also matter. In late colonial Angola, the EOA broadcast in FM, but all the other radio stations—clubs, commercial, private, and religious—broadcast in shortwave and/or medium wave.

Data on radio range and reception is essential to understanding radio's work and power, and though insufficient, I used it when I could find it. Radio's magic and charm come from the gap between what science can explain, fluctuating atmospheric conditions and physical obstructions, and how human perception operates. This book happily inhabits that murky terrain. It rests in the folded wire and beads of the artist's radio and the way it represents the aspirations to modernity, the desire to embellish the machines of everyday life with imagination, and the fact that it is a working radio, albeit with limited reception.

White settlers in Angola thought about broadcast range and reception. They thought about those questions across overlapping local, regional, and imperial maps. Chapter 1 lays out the early days of amateur and then colonial state broadcasting in Angola. I explore the relationship among white settler identity, the occupation of sonic space, and whiteness between the 1930s and 1970s. This is a world regularly invoked in former settlers' nostalgic reveries of the colonial period. One woman, who was born and raised in Angola and then left as a young adult but continues to identify as Angolan, told me, “Radio was our oxygen,” leaning on metaphor to express why and how radio’s immateriality breathed life into a white world on the edge of empire.

1 ⇜ Sonic Colony

Whiteness, Fast Cars, and Modernity, 1931–74

THE FIRST radio transmission in Angola took place 28 February 1931 in Benguela, a city on Angola's southern coast, from the CR6-AA Broadcasting Post of Álvaro de Carvalho.[1] Angolans offer this detail with empiricist delight because this broadcast takes radio history away from Luanda, thus decentering and deinstitutionalizing it. Throughout this chapter, I illustrate that Angolan radio originates from within the colonial settler population and not the Portuguese colonial state. Angolan radio's creation narrative is also the story of the colonial Angolan soundscape as the foundation for creating and structuring white settler life and identity.

The entryway into my narrative about the sonic construction of white settler identity—and the idea of sonic colony—is "Paraíso" (Paradise), the second half of Miguel Gomes's feature film *Tabu* (2012).[2] Structured around a tragic colonial subject, Aurora, and her illicit romance with Gianluca, the Portuguese director fashions a story about an insulated settler world in the final stages of the colonial period.

Set in a Lusophone African "somewhere," at some point before, but near, the onset of the African wars for independence from Portugal, "Paraíso" opens with a sequence of shots establishing Aurora's comfortable colonial life. Viewers learn from Gianluca's voice-over narration that Aurora's mother died giving birth to her and that her father, despising the monarchy, resettled in Portuguese Africa. Though he made a fortune in textiles, he lost much of it to a gambling habit (like the one that ruins Aurora's later life in the film's first part). Nonetheless, upon her father's death, Aurora—haughty, imperious, reckless, and self-destructive—finds her way to university, marries a wealthy tea farmer, and lives on a large estate at the foot of Mont Tabu.

Against the narrator's background detailing, Gomes presents images of Aurora's leisure: she practices sketching (with mediocre results); she tutors a young African girl in writing Portuguese; from a veranda, she oversees a tree's pruning—extending the view of her landholdings. We also watch as Aurora stretches, warming up before taking a jog, ostensibly around these grounds. Over this last shot—Aurora running out the open door of her manse into a bright white opening as two African servants sweep clean the front parlor—the title card appears and holds on-screen: "segunda parte: Paraíso" (second part: Paradise).

Freezing the film here, the shot fascinates for what it expresses in a latent, though startling manner about Gomes's narrative. Ruí Poças, the cinematographer, sets up this image as a sequence of three descending frames. Against the whiteness of the central, narrow frame, note the two African servants, silent in their labor expect for the shushing of their straw brooms, one farther, one closer. Bright light darkens the corners of the frame, erasing the details of the servants' faces and casting their figures in outline. The silent, almost invisible, labor of African servants supports white leisure and levity.

In that shot, darkness obscures the foregrounded African servants (fig. 1.1). More typically in "Paraíso," Gomes presses the Africans into the background, flattening and silencing them. In both cases, reminiscent of the cutout silhouettes in the works of the African American visual artist Kara Walker, Gomes and Poças render the Africans as one-dimensional paper-doll cutouts, replicable and disposable. They nod and move stoically around the gaiety and tragedy of European settler life as it tingles and stings, sparkles and galls, whispers and rocks out. Like children in the old adage, Africans in this colonial story are best seen and not heard: European civilization has tamed these Africans and the savage darkness they embody, made them obedient, tucked them into starched uniforms, and silenced them.

Actually, with the exception of Gianluca's extended memory and diegetic sounds (from nature, African ritual, or popular music), Gomes silences everyone. Though viewers see the white characters' mouths moving as if in dialogue, we do not hear them at all. Alternatively, the Africans appear voiceless, mute. Who has silenced them, stolen their voices? Does Gomes want viewers to imagine that the settlers and the colonial state are deaf to African humanity and to African realities?

FIGURE 1.1 A still from Miguel Gomes's 2012 film, *Tabu*, that captures the relationship between white and black, light and dark in an unnamed Portuguese colony in Africa. Used with permission of O Som e a Fúria, 2012.

Here I'm invoking Mara Mills's liberatory conception of deafness as "a variety of hearing; alternately, it can be conceived as a precondition of hearing or as resistance to hearing and audism."[3] In *Tabu* deafness isn't liberatory; it's a willed refusal to hear (on the part of an invisible state) and the privileged, cultivated ignorance and distance of the settler. Considering the links between whiteness and power, "deafness" and "muteness" are crucial metaphors. Deafness describes the colonial state and settlers' refusal to hear Africans, while muteness denotes how Africans are silenced, their voices unimaginable in the sonic colony.[4]

Gianluca acknowledges this when he states that the white settlers of his "quiet" part of the colony disbelieved the initial rumors of Africans raising arms against and massacring whites in other settlements: "In the quietness of our community, the news sounded exaggerated, even unreal." Refusal to hear or believe Africans is a precondition of Romantic insularity. The invention or maintenance of Romantic adventure in exoticized southern Africa is predicated on colonial violence and the exploitation that allows for the fantasy to flourish. "Community" in Gianluca's telling means whites only; the fantasy of tamed, open, undiscovered savannah necessitates excluding black Africans—with their ideas, desires, and their capacity to express them—from the settlement dream. They could only serve.

As the narrator's voice-over continues, Gomes links a sequence of shots framing groups of silent southern African men, women, and children staring intensely into the camera (a rare scene where Africans are not in the background but front and center; see fig. 1.2). Gomes shows their silence, produced in the act of colonization. Viewers see some of these people looking back. The tension between the silence and returning the camera's gaze presses on the viewer's conscience, making us wonder what alternative stories these people might tell.

Once the news of revolt and killing is confirmed, Gianluca continues, "white folks got excited and organized militias, night patrols, and did shooting exercises accompanied by tea and biscuits." Viewers hear these practice shots ringing out as they see puffs of straw dust dart out from the torsos of dummy targets. The sound of the gunshots anchors the image and the colonizing gaze itself. The sound reminds us of the camera's association with early gun technology, hand-cranked cameras and machine guns, both crucial to the colonial project.[5] Tea and biscuits are predicated on the gun, on latent and active violence.

Since hearing is material, sensuous, and rooted in the body, when Gomes allows his audience to hear those shots, he's forcing us to experience ourselves in the settlers' world. In her phenomenological reading of whiteness, Sara Ahmed suggests that this sensuous experience—to which I am adding the auditory—is part of what makes "whiteness worldly."[6] The sounds and silences, as much as the image's frame, situate the viewer. We hear with white ears.[7]

In Portuguese colonial Africa, whiteness was loud, bold, and dangerous, full of possibilities not available in the gray, whispering, careful Portugal of the Salazar and Caetano regimes. "Paraíso" evokes the world that settler broadcasting built in Angola. Settler broadcasting in Angola linked and built, rocked and fado-ed, semba-ed, and spoke Umbundu even as it sounded out a Portuguese Africa. This was the foundation of all broadcasting and listening culture in Angola. As I turn to discuss early Angolan radio, it's this auditory phenomenological white experience that I want to amplify, clarify, and define in thinking about radio sound as world-constructing.

This is what I mean by "sonic colony": a colonial space defined and permeated by sound. What did the Angolan settler colony sound like? How was sound colonizing? Sound varies over time and space. Gomes

FIGURE 1.2 A series of shots from *Tabu*. These scenes underscore quietness and muted African voices as the grounds on which colonial settler society is built. Used with permission of O Som e a Fúria 2012.

In the quietness of our community

the news sounded exaggerated, even
unreal.

The confirmation arrived through a military officer

uses sound to give viewers the experience of the polarities colonialism produced and its sensuous elements. He introduces radio briefly, subtly, to alert us to affect and technology, sound, sadness, and nostalgia. Sitting with the radio just behind and to her side, Aurora dips into silent, melancholic reverie about her lover as Mickey Gilley's "Lonely Wine" plays on the radio and her friends are bound in joyful camaraderie. At a later point, she sobs in counterpoint to the upbeat sounds of the Madagascan band Les Surfs' rendition in Spanish of "Be My Baby." The sounds on the radio move Aurora into her sadness.

In the radio sounds Angolan settlers produced, we can find a set of practices and sensibilities that help us understand some of the processes of late colonialism in the territory. Sound held the settler colony, if not the colony as a whole, together. Fourteen times the size of Portugal, when electricity that clattered through the air could be captured to bring one city and town closer to another and all of them closer to the metropole, settlers seized the opportunity to harness it. Radio enriched daily life and gave settlers a way to be modern, more modern than the Portuguese in Europe. In this sense, radio was oxygen—life-giving, energizing, an invisible connection to other places.

RADIO AND MODERNITY

In other colonies on the African continent, European populations also introduced radio—in the Gold Coast and South Africa, for example.[8] The colonial state took over broadcasting much earlier in these places. In the Portuguese colonies, settlers remained the primary radio broadcasters for a much longer period of time. Little discussion of those European broadcasters looks at what it meant to them or how to think about radio, modernity, and whiteness. Scholarship on radio's modernity focuses on how the colonial state and the postcolonial state used radio as a symbol of modernity.

Radio projected modernity and the civilizing presence of the colonial state. The history of radio on the African continent exposes how colonial states introduced radio to cater to European colonial administrators and settlers and, later, as part of developmental colonialism. The radio is a piece of modern technology that uses electromagnetic sound waves to overcome distance. Radio stations are modern institutions that marry technology, technical knowledge, and the rationalization of labor. As part of colonial rule, Brian Larkin, writing on Northern Nigeria in the 1950s, observes that "radio materialized a theory of rule and administration, a new relationship between the state and its subjects."[9] This theory rooted itself in a belief that modern technologies are, ipso facto, modernizing agents. Debra Spitulnik, analyzing the "saucepan special" in Northern Rhodesia, calls radios "mobile machines" in order to capture the circulation of these objects dense with significance.[10] Radios bore status, modernity, prestige, and communicative frisson.

Advertisements for radios directed to African audiences addressed them as modern consumers.[11] The saucepan special—large, enamel, and unwieldy—fit for what colonialists deemed nonnimble African hands and woolly heads, took off in the mining towns of Northern Rhodesia. The transistor radio revolutionized listening in the 1960s and 1970s and opened space for Africans to engage in acts of self-styling with the technology. Small, mobile, battery-operated, and inexpensive, transistor radios made shortwave stations accessible to thousands more listeners. Mhoze Chikowero captures the radio's many valences: "In colonial Africa, [the radio] was simultaneously a dubious tool of colonial assault, a new, fascinating communication and recreational technology, a prized, symbolically powerful piece of property and a weapon that enabled the colonized to engage the colonist."[12]

Colonial powers, exercising a kind of technopolitics, employed these technological innovations to promote state policy and project imperial authority.[13] Radio infrastructure and programming promoted civilizational politics.[14] But against what Larkin, Spitulnik, and Chikowero argue for Nigeria, Northern Rhodesia, and Southern Rhodesia, it was not until the mid-1960s (by which point Nigeria was independent and Southern Rhodesia was the renegade white-settler-ruled state of Rhodesia), when the colonial state in Angola broke ground on a new radio station, that state radio and modernity officially intertwined. From its first introduction in 1930, the relationship between state and radio, as an object and as an institution, has been more tenuous than in those other colonies.[15] Instead, radio's modernity resonated with the growing white settler world.

Before we can think about how the late colonial state and, later, the independent Angolan state used radio to consolidate and extend power, we need to consider its very unstatelike origins in amateur radio clubs. The white settler network of radio clubs shaped how people inside the territory listened, as well as the sounds of place and broadcasting, and permeated how settlers understood themselves. Gomes's film offers a fictionalized account of such a scenario. The white radio networks reverberated in Portugal after Angolan independence when settler broadcasters moved there and reimagined communication for a democratic Portugal. In other words, the modernity associated with radio and its sound impacted not just African subjects but also the white settlers sent to exemplify civilization and materialize occupation. That produced feedback in Portugal, too.

Like Algeria, South Africa, Kenya, and Southern Rhodesia, Angola was a white settler colony.[16] Though the extreme segregation that characterized South Africa and Southern Rhodesia never existed in Angola, settlers practiced, and the late colonial state encouraged, residential, occupational, and educational segregation. Racial barriers structured late colonialism.[17] Unlike in South Africa and Rhodesia after World War II, Angola did not have a color bar, because Africans there had never possessed enough economic power to threaten that of the white settler population.[18] Nonetheless, a surge in white settlement intensified spatial, professional, and intellectual divisions in the second half of the twentieth century. New laws strengthened state administration and reified racialized rule.

These changes in governance started earlier in the century. Similar to other colonial regimes, the Portuguese extended control and administration of populations through taxation and pass laws in the first part of the twentieth century.[19] The *cadernata,* at first a pass for "natives," later became the basic identity document, while the "civilized" (*assimilados*) boasted a *bilhete de identidade* (identity card).[20] Laws promulgated in the 1920s and 1930s created a bifurcated and unequal society. They followed the segregationist policies in French and English colonies:

> The 1926 "Civil and Criminal Statute for the Natives of Angola and Mozambique" summarized practices and sparse colonial legislation, creating a special legal status for "natives" distinct from "Portuguese citizens" and impacting on all related legislation. It broke definitely with more liberal tendencies and matched segregation reinforcement in other African colonies. It was a convenient barrier against black social mobility in white settler colonies, protecting settlers' rights to land and jobs from the overwhelming black majority, especially the small but growing group of Christian mission trained people.[21]

The divisions between "natives" and "civilized," which included all white Portuguese, protected white settler access to land, wealth, rights, and employment. Racism, not nationalism, provided the glue for settler identities, pushing Portuguese and other whites together.[22] It makes sense, then, to think of Angola as a white settler colony.

The historical literature on European settlers on the African continent, particularly on Angola and Mozambique, is growing.[23] Luise White rejects "settler colonialism" as a term for her study of Rhodesia in order to highlight the significance of African decolonization, of a politics of sovereignty surrounding Rhodesia from the outside rather than a racial politics dividing it from the inside.[24] That is an important and eye-opening project. But for the Portuguese colonies in Africa, using the term "settler colonialism" can rend the ideological stranglehold of "lusotropicalism"; it can allow us to hear the construction of an insular world amid the claims about racial mixing and approximation that the metropole used to justify its continued presence on the continent. Pressing forward, comparative settler history, or at least its invocation, trains our ear to the white political parties whose volume MPLA nationalist history has deafened us to, in order to amplify the party's nonracialism. Comparative settler history can help us understand ALCORA, the secret agreement between South Africa, Rhodesia, and Portugal, to collaborate and stem the tide of black rule in southern Africa.[25]

The Portuguese state promoted settlement as a strategy at various times across Angola's colonial history to vent the criminally and politically condemned, to combat the depredations of the slave trade, and to demonstrate occupation in the wake of the Berlin Conference. But Portuguese immigrants never thought of Angola as a priority destination until after World War II, preferring Brazil or European destinations.[26] When the state replaced the 1951 Colonial Act with the Organic Overseas Law in 1953, it tightened the legal embrace between Portugal and its African colonies but did nothing to pull real lives closer together. Indeed, settlers in Angola, like those elsewhere, found themselves at odds with the metropole, the colonial state, and Africans.[27] Though the Overseas Law created a new legal status for the colonies and introduced the language of development, policy implementation lagged until the anticolonial war broke out in 1961.[28]

The *colonatos* (white settlements) at Cela and Matala in the 1950s, a product of this thinking, suffered from low numbers of settlers, costly expenditures, and the resentment caused when the state displaced African populations from the lands they inhabited.[29] Attempts to integrate the colonatos, after the war erupted in 1961, likewise faltered.[30]

Other Portuguese migrated but did not settle in the colonatos. When the colonial war catalyzed the development of local infrastructure and the economy, many educated Portuguese saw opportunities unavailable in Portugal and in Europe.[31] Higher numbers of professionals migrated, even as the number of Portuguese engaged in petty commerce grew steadily.[32] For the late colonial period (1940–75), Cláudia Castelo concludes that contra Gerald Bender, the average settler enjoyed a higher level of education than the average Portuguese in the metropole or those who immigrated to other European countries. Greater gender parity reigned (relative to the very skewed ratios of earlier centuries), creating more insular colonial communities.[33] The Salazar state thus succeeded in creating migratory flows motivated by incentives other than poverty.[34]

For the Estado Novo, avoiding white poverty in the colonies took priority. They sought educated and skilled migrants to help build the industrializing economy and to staff the colonial administration.[35] Settlers created a colonial lifestyle, building the ideological edifice of empire, extending the veil of social and cultural practice that would divide "civilization" from its "Other."[36] As in other white settler colonies, whiteness became the lowest common denominator, the edge over which not to drop, what David Roediger called the "wages of whiteness" in the nineteenth-century United States.[37] White poverty endangered the idea of European superiority, whether it be in Kenya, Rhodesia, South Africa, or Angola, the US or Australia, so states implemented color bars and white welfare to engineer difference.[38]

Large white populations pooled in urban areas, despite attempts to settle Portuguese in rural colonatos.[39] Settlers, after all, pursued their own interests and had their own ideas about the good life. New settlers saw the opportunity to create a modernity unavailable in the metropole, where the grip of the Estado Novo, its conservative Catholicism and dire fascism, suffocated freedom. Some had sought political refuge in the colonies for quite some time (like Aurora's father in Gomes's *Tabu* film). Many pursued uncensored cultural space. After World War II, and with new investments and the economic growth stimulated by the war, migration motivated by pull factors reached its height.[40] White urban populations grew. Castelo notes:

> If in the '20s and '30s the principal cities of Angola and Mozambique were nothing more than provincial towns, without basic sanitation or electricity, from the '50s they became progressive, modern areas with a dynamic social, cultural and recreational life, and relatively liberal in terms of public opinion and in terms of habits, within the limits of censorship and racial barriers.[41]

This late colonial immigration struck a contradictory chord. At the very moment Salazar sent Freyre to the Portuguese African colonies, the numbers of white settlers peaked, sharpening racial divisions.[42] Portuguese rule in Africa faced international hostility in a world where independent former colonies constituted a majority at the United Nations. In the end, immigration exacerbated racial divisions. Even Freyre agreed.[43]

Nonetheless, Angola's white settler communities offered an alternative world for white Portuguese, if not a salvation for Portuguese diplomacy. By 1970, permanent white residents made up 5.1 percent of the total population, the second-highest white settler population on the continent after South Africa.[44] Many of them were recent arrivals.[45] Southern Angolan cities held the largest per capita Portuguese populations. By 1970, the city of Sá de Bandeira (Lubango) had a majority white population. It was the site of the most dynamic radio scene. First, some thinking about whiteness in Angola before circling back to Sá de Bandeira, its settlers, and the radio.

BECOMING WHITE

The settler strongholds of the central and southern cities distinguished themselves in the realm of radio, aviation clubs, and motorsporting. White settlers used sound and speed to assert their modernity, to overcome second-class whiteness, in some cases fostering an idea of white Angolanness and white politics.[46] The growth of separatist white politics in these areas in the 1960s was no mere coincidence.[47] The advent of independence stymied the realization of separatist and other white-identified politics.[48] But radio, memory, and political affiliation in Portugal and in Angola after 1974 still reverberate with the effects of these earlier cultural practices and politics. How did modernity, sound, and speed foster whiteness or becoming white?

The Portuguese state promoted whiteness, occupying space through settlement and through the segregationist laws mentioned earlier. Laws linked civilization, citizenship, and whiteness, naturalizing their association. The Estatuto Indígena (Native Statute) defined two groups: "civilizados" (civilized) and "indígenas" (native/indigenous). The state classified all whites as "civilized" by default. A small number of black and mixed-race individuals could qualify if approved by colonial authorities, who evaluated their religion, home, language, educational achievements, and the like. The civilizados who were black or mixed-race (later called *assimilados* or "assimilated") never constituted even 1 percent of the population, an indictment of Portuguese rule on its own terms.

Castelo and Maria Conceição Neto argue that the Estatuto Indígena structured the colonial world. It arranged material space and shaped human experience. Sara Ahmed argues that whiteness is phenomenological, a social skin that creates space and the experience of it for whites and nonwhites.[49] In her words:

> Colonialism makes the world "white," which is of course a world "ready" for certain kinds of bodies, as a world that puts certain objects within their reach. . . . Whiteness is an orientation that puts certain things within reach. By objects, we would include not just physical objects, but also styles, capacities, aspirations, techniques, habits. Race becomes, in this model, a question of what is within reach, what is available to perceive and to do "things" with.[50]

Portuguese settlers in Angola reached for modernity in radio, motorcars, and small planes—not to mention the very idea of civilization.[51] Whiteness as an orientation made using these objects and ideas natural, comfortable, and unself-conscious.

Sound requires its own kind of orienting. One turns to the radio, turns attention to some sounds and not others. The cultural worlds that settlers in Angola built around radio, aviation, and motorsporting marked their difference from Africans, from Portuguese in the metropole, and brought them closer to other Europeans in the region. Turning to the radio, they extended whiteness into the soundscape and produced social comfort sonically. Gonçalo M. Tavares writes about the moral quality of listening. Remember Gomes's film as you read this:

> Tell me what you hear, I will tell you the weight of your guilt. What sentences you hear, to what words you give attention. What music you listen to, to what music you give attention . . . Deciding which sound to give your attention to is, in the end, to decide at what volume you will place the different parts of the world. It's an important decision.[52]

For Ahmed, these orientations are less volitional than Tavares makes them out to be. The way whiteness worked in late colonial Angola oriented Portuguese settlers to certain sounds. New technologies reverberated with modernity. Whiteness put them within reach. Settlers used them to define a whiteness that arced away from continental Portugal. White settlers made decisions about where to turn their ears in attention and where to direct the sounds they produced, whether locally, in the region, or back to the metropole.

De jure definitions of whiteness afforded protection and put things in reach. It remained to settlers to delineate the practices and contents of whiteness and to become white through social and cultural practice. Portugal constituted part of Europe's southern rim. Boaventura de Sousa

Santos charaterizes Portuguese colonization as peripheral and describes the Portuguese as "colonized colonizers."[53] Subaltern to European norms (and particularly those of the United Kingdom), the Portuguese used their relationship to Africa and Africans to erase their liabilities vis-à-vis Europe. Portuguese thus defined their whiteness and modernity in two directions, "as Prospero reflected in Caliban's mirror, as Caliban reflected in Prospero's mirror."[54] Santos points out that Europeans applied the same stereotypes used to disparage peoples of the Americas and Africa to the Portuguese.[55] Jason Keith Fernandes puts a finer point on it: "Even if Portugal may have been seen as European, not all Portuguese were seen as participants in European culture."[56] The Portuguese had to become white to become European, in other words. Settlement provided a way for both continental and settler Portuguese to orient themselves within Europe and against Africans. The Portuguese state asserted Europeanness through colonization, whitening its own population through political rule. In the colonies, settlers became white relative to black African and mixed-race peoples through cultural practices, like radio and motorsports, but they also distinguished themselves from the metropole and came to identify with Angola.[57]

Whiteness in the colonies was not monolithic. The law categorized Portuguese born in Angola as "Euro-Africans," what people referred to in colloquial terms as *brancos de segunda*, or second-class whites.[58] This status and naming had material and cultural effects on the lives of those Portuguese born in the territory. For example, the Official Angolan Broadcaster (EOA; 1953–74) employed primarily Portuguese born in Portugal. Those born in the colony entered public service only with the help of insiders.[59] These hierarchies shaped whiteness in the colony and framed work in radio. Angolan whiteness, modernity, and radio overlapped.

The particularities of this legal and cultural mix, of this status, and the cultural worlds that emerged through it sometimes radicalized Angolan-born whites.[60] Many of those who moved there while young, or were born in Angola, quickly identified with the space and its sounds, smells, and land.[61] They spoke Angolan Portuguese. The children of *fubeiros* (the owners of shops in the informal mostly African neighborhoods, called *musseques*, and in the interior) grew up playing with black and mixed-race children.[62] The colonial state organized the economy to elevate settlers over Africans and the metropolitan elite over settlers. By the 1930s, some settlers began to protest, translating economic marginalization, distance from the metropole, and approximation to Angola into political difference.[63]

Sound was a key part of this differentiation. What bound sound and whiteness? Let's return to that first broadcast from Benguela by Álvaro de

Carvalho in 1931. He played Cruz e Sousa's march "Feno de Portugal" and then the *fado* (Portuguese folk song) "A Tendinha," sung by Hermínia Silva. These two pieces epitomized Portugal under the Estado Novo: a military march and a fado, quintessentially modern in production technique and style, thoroughly nostalgic in content, one upbeat and martial in tone; the other melancholic and full of suffering.[64] Gabriela Cruz argues that the Estado Novo promoted new media technologies (radio, cinema, telephones, and turntables) in the Portugal of the 1930s and 1940s that "drastically redrew the terms of sensorial experience, making way for the emergence of new modes of perception and enjoyment, which were politically consequent."[65] The technical qualities of sound reproduction in film created a sense of electronic suspension of fado and of vocal purity.[66] Mass culture spread modern media forms that circulated this content.[67] Modern media, John Durham Peters observes, are instances of "applied physiology" that reroute the senses, the human body, and the national body.[68] Technology does not determine, but it has effects. We understand differently not just because we have ingested new content but because technologies transform our physical dispositions and our consciousness, our ways of perceiving ourselves and the world around us. They orient us.

In Angola, modern technologies "redrew sensorial experience," but with a difference. To begin with, the Estado Novo invested little in the territory. The bureaucratic technologies of the state and practices of national (Portuguese) identification attenuated there. Settlers, more than the state, used technology to connect and communicate. Settlers used radio and aviation to facilitate contact, to extend the range of ideas circulating in the locally owned press.[69] Diamang, the diamond concession company, recorded Tchokwe musical forms threatened by Diamang's mining, capturing disappearing practices for study and museum display in what Delinda Collier refers to as "information colonialism."[70] Technologies thus mediated Portuguese settlers' sensorial experience of Angola. At the same time, the senses worked as a technology, a way to understand affective responses to settlement in a distant land. Castelo notes that "vision, listening, smell and touch are called upon to translate the environment," and thus the language of the senses permeated settler writing on Angola.[71]

Put differently, settlers entered this distinct sensorial world and remade parts of it through modern technologies like radio. That fostered an alternative politics. The size of Angola, the distance between areas of white settlement, and the desire to connect them motivated the use of technologies. Radio did not just link Portuguese communities, it helped "extend the skin of the social" that, following Ahmed, created whiteness as a space of comfort.

If, in the Portugal of the 1930s and 1940s, following Cruz, new media technologies and the Estado Novo fostered a sense of transparency and affective interiority, in Angola settlers used radio to produce an "Angola" structured through and in sound, an insular—if diversely sourced—white world, a narrow "we" in a distant land, distinguished from metropolitan Portugal in its embrace of the self-conscious speed and harnessing of nature that characterize modernity. Radio broadcasted whiteness, stretching it out sonically, hailing listeners to orientate themselves around and to the sounds of settler whiteness.

SÁ DE BANDEIRA

This section looks at one particular area: Huíla, where the capital city, Sá de Bandeira, had the largest per capita white population by 1970 and had become the communications hub connecting the southern region of the country to the northern and central regions.[72] In 1970, Sá de Bandeira's population had a white majority, the only such city in the territory. Whereas the last section focused on settlement from the 1950s on, here I will press back a bit earlier. The city of Sá de Bandeira took its name from the nineteenth-century governor of the colony, Marques Sá de Bandeira, who believed that white settlement would allow the colony to prosper without the slave trade.[73] The city named after him was meant to be such a place: a bastion of productive white farmers in the Angolan interior. An initial attempt at the settlement of two hundred whites in the late 1850s failed, in part due to poor communications with and transportation to the port city of Moçamedes.[74] The metropolitan state obliged settlers to build this infrastructure (*CH*, 157). It should come as no surprise that when settlers wanted to improve communications, they did so themselves by building radio clubs.

Southern Angola was a late addition to the Portuguese Empire. The staggered settlement of different empires in southern Africa trammeled southern Angola. Attempts at settling the interior included those by Boers, from South Africa's Transvaal, and by Madeirans in the 1880s (*CH*, 119–267).[75] Boers fled British rule in South Africa. They settled in Humpata in the early 1880s and helped conquer the Huíla highlands for the Portuguese, violently dislocating Nyaneka-Humbe, providing transport, and opening roads (*CH*, 164–88). Their temporary stay opened the area for Madeiran settlement (*CH*, 188). From 1884, the year the Berlin Conference divided up the African continent, ignoring African political rulers and borders and demanding that European powers demonstrate "effective occupation," settlement in Sá de Bandeira, and in Huíla in general, grew in tension with the Portuguese state

(*CH*, 137). The state's financial support never matched its rhetorical flurry. The first group of Madeiran settlers (noncontinental Portuguese) arrived in 1884. They encountered challenging material conditions and insufficient support from the colonial government that had contracted them (*CH*, 195–96). Historical accounts are rife with tales of arduous journeys: pioneer columns suffering hypothermia, food shortages, and deaths along the trail from the coast to the interior in rented Boer wagons. A second contingent of 335 Madeiran settlers arrived in August. Between 1884 and 1892, a total of 2,075 Madeirans settled in the area, founding the colony of Sá de Bandeira and increasing the white populations in Chibia and Humpata (*CH*, 201–3, tables).

A temperate climate would foster settlement, the thinking went, but the land was not particularly fertile except along waterways (*CH*, 86, 89, 97, 157–60). Adapting to these conditions, Nyaneka-Humbe populations and Boer settlers raised livestock (*CH*, 97). Portuguese settlers struggled. Madeirans, Clarence-Smith tells us, "were desperate poor small farmers, often illiterate, who were accustomed to intensive irrigated agriculture on tiny plots of land in their native island."[76] Carlos Medeiros notes that colonial administrators criticized their laziness, lack of agricultural experience, and drunkenness but concludes that these characteristics held less weight than administrative, infrastructural, and organizational problems (*CH*, 211–12). The Portuguese assumed that their stereotype of Madeirans—that they moved as families and set down roots—would substitute for material support (*CH*, 190). In the end, Madeiran settlers exhausted the soil, subdivided the land, and fell into debt. Some were forced off the land to become sharecroppers or wage laborers. Poor transportation and communications infrastructure (or its complete absence) created an isolated, insulated area, complicating economic exchange throughout the territory and region (*CH*, 59, 213–16, 253–66). For those who held on, this became an intensely inward-looking community: hardworking, suspicious of outsiders, bound to the land and their cattle.[77]

The colonial government declared Sá de Bandeira the capital of the district in 1901, after which the town profited from state favoritism, a larger settler population, and judicial, administrative, and military offices (*CH*, 243–44). In World War I, the colonial administration and military began to develop highways and railroads in the topographically challenging region, yet the state completed asphalting the main artery to Luanda only in the late 1960s and other regional roads in the early 1970s (*CH*, 258–59).[78] High Commissioner Norton de Matos expanded transport infrastructure and prioritized settlement and settlers in the early 1920s. In 1928, settlers from northeastern Portugal (Trás os Montes) received state support in the form of transport, less regulation, and access to larger areas of land (*CH*, 272–87, 294). By the late 1920s, their ranks swelled by demobilized soldiers (fighting the Germans in South-West Africa) and a trickle of other Portuguese immigrants, the white Portuguese population in the Huíla highlands reached nearly four thousand (*CH*, 258). Smallholder subsistence agriculture, civil service work, and small commerce dominated the economy (*CH*, 247–48).[79]

A small white elite of heavy-handed capitalists dominated southern Angolan life from the 1920s on. Like the rest of Angola, Portuguese migration swelled the city of Sá de Bandeira beginning in the mid-1940s such that by 1970 Castelo shows the white population at 13,429.[80] Changes in education introduced after 1961 desegregated and extended state education. In Huíla this meant that once poor whites had greater access to schools and white and black students would have begun to mix at the elite high schools (*liceus*) in Luanda and Lubango.[81] The children of political dissidents rubbed shoulders with the sons, grandsons, and sometimes daughters of settlers born in the land and others more newly arrived at the city's elite secondary school, the Liceu Diogo Cão.[82]

Whites constituted most of the petty bourgeoisie in the city and the region (something not true for the rest of Angola).[83] Whiteness provided a buffer against the African population: "One visitor after the First World War commented on the fact that Moçamedes did not have the 'trousered blacks,' who were such a prominent feature of Luanda and Benguela. The existence of a significant stratum of poor whites in the south had much to do with this anomaly."[84] The region's whiteness, and its distance from the seats of power in Luanda and Lisbon (its far southern location, Cuando Cubango, drawn for the Portuguese imagination in the phrase "the ends of the earth"), marked its specificities. Clarence-Smith and Medeiros underline the lack and difficulty of transport and communications in the region.[85] But by the early twentieth century, some settlers thrived. Pimenta discusses Venâncio Henriques Guimarães and the economic empire of cotton, cattle, and commercial exchange he had developed in Huíla and Moçâmedes by the 1920s.[86] Guimarães and other successful farmers and businessmen founded numerous commercial and agricultural associations to defend their interests in central and southern Angola.[87] These individuals and associations fought for colonial reforms and for greater settler autonomy in the 1960s, but their business interests led their political ideology.[88] Only a few, such as Fernando Falcão, leader of the FUA (Angolan United Front), addressed explicitly political issues since such concerns got little quarter from the metropolitan state.[89]

With Portuguese immigration increasing, Lubango became a key communications and administrative center in southern Angola by 1960.[90] Its whiteness and its distance from other settlements spurred the desire to be connected and to be a hub for telecommunications, transport, and bureaucracy for this late colonial generation of settlers. Into the 1970s, Sá de Bandeira was the largest and whitest of Angola's urban spaces.[91] Ana Sofia Fonseca's *Angola, terra prometida* (Angola, promised land) reconstructs the city of the late 1960s and early 1970s from the memories of her informants: "Sá de Bandeira is, clearly, a prosperous city, bustling and white. The great majority of blacks live outside its best perimeter."[92] For the white youth of the day, "the PIDE [secret police] doesn't bother them, the war is distant, and the reality of blacks is almost non-existent."[93] This insular if outward-looking community, perched on the edge of the Portuguese Empire, generated powerful sounds.

RADIO CLUBS

Angola Radio, a monthly publication dedicated to radio hobbyists, debuted in February 1941. It ran for a year. The magazine's cover and contents expressed the idea of radio as modernity. A graphic design of a radio tower superimposed on a river evokes Heidegger's sense of the productivity technology harnesses, or challenges forth, from raw nature.[94] Another cover shows a nattily clad white couple in the embrace of a social dance, drawn over the tower, pulling modern entertainment down from the ether (see fig. 1.3).[95] Between the magazine's covers, local radio producers asked the state to recognize them as key nodes of imperial propaganda. Published in Luanda, the magazine represented a set of interests somewhat different from those that emerged around radio clubs based in smaller cities. It turned more decidedly to the metropole, a position evident in editorials. Perhaps this metropolitan focus accounted for the publication's short run.

The August 1941 issue featured an editorial titled "Radio at the Service of Empire" that advocated stronger broadcasting from Lisbon to combat "the constant beating on the door" of foreign voices.[96] A response in the next month's issue echoed this concern but suggested that Lisbon devolve the work to local radio clubs and provide them with resources.[97] Regime supporter and propagandist Reis Ventura chimed in in October's issue, praising "radio and aviation for abolishing physical and sentimental distance" transforming once distant, indistinct places into "nearby and fertile lands."[98] Radio and air transport made the empire's periphery accessible and productive. Radio sound shrank the gaps between Portuguese settlements. It pulled the metropole

FIGURE 1.3 The cover of *Angola Radio,* July 1941, showing the association between whiteness, leisure, social dance, and modernity through radio. Used with permission of Rádio e Televisão de Portugal (RTP), Área Museulógica e Documental.

nearer. Amateur radio clubs mushroomed around the settler regions of the country's central and southern highlands. New technologies linked them across space and time, pulling the whiteness of modernity's electric buzz down from the ether to outline these communities in sound.

Álvaro de Carvalho's first radio transmission from Benguela occurred some 360 kilometers from Lubango/Sá de Bandeira. Radio enthusiasts at the Rádio Clube de Huíla made their first experimental broadcasts in May 1939 and officially inaugurated the club in October of the same year with a 50 kW transmitter, a pickup disc player, some LPs, and a microphone. Initiated around the same time as the Rádio Clube de Angola, based in Luanda, they oriented themselves to the Rádio Clube do Sul—based in Benguela and Lobito—the outgrowth of Carvalho's home station. A greater density of clubs existed in the territory's southern region. By the mid-1950s there were a total of fifteen broadcasters—mostly radio clubs but also Diamang's private station, the state broadcaster, and Rádio Ecclesia.[99]

White settler men dominated in the radio clubs. Like the commercial and agricultural associations in the area, local elites formed associations, paid monthly fees, purchased equipment, and set up studios.[100] In Lobito senior employees of the port founded a radio club, and in other cities radio club membership typically included local doctors, engineers, army captains, and often the municipal governor. All the clubs constructed buildings. The number of clubs grew in the 1940s, and by the 1960s they had become a firmly entrenched phenomenon. These men (and some few women) were radio amateurs and radio enthusiasts, not trained journalists. In a world without journalism schools, each employee did a bit of everything. Only the metropole knew the luxury of specialization.[101]

Carvalho's first transmission evoked a strong identification with Portugal as a homeland. The Huambo Radio Club's call sign proclaimed "a Portuguese voice in Africa." Eventually radio in Angola created an Angolan sound. Later radio enthusiasts sonically declared Angola their homeland. Men like Sebastião Coelho, Leston Bandeira, David Borges, and Fernando Alves incorporated African languages, the burgeoning Angolan urban popular music, and the bonds of Portuguese community they had developed in situ into their radio programs, whether music shows or sports reporting. Their memories reconstruct a world of freedom, male sexual liberty, and modernity on the edges of empire. Associated with it were the sounds and smells of local flora and the laughter of friends and neighbors, creating a soundscape dense with significance beyond what the metropole provided.[102] A small industry of memory and history of settler radio attests its continuing significance in people's lives.[103]

Domingos Peres founded the Huíla Radio Club in 1938 and it started broadcasting from the home of Cândido Alves Espinha. The club had a six-person directorship that elaborated the club's statutes, which guided the activities of more than two hundred members. With a 50-watt transmitter, they operated with as much reach as the larger Luanda-based Angola Radio Club. In 1944, with some state support, they broke ground on a building, and by 1948 they had purchased a 1 kW transmitter.[104] A dozen years or so after its founding, the club boasted eight hundred members and a library that held three thousand records. The club merged with the Angola Aero Club in the mid-1950s, and together they built a headquarters, suggesting continued growth and the interest in increasing air-linked travel and communication.[105]

A collection of documents in the António Salazar archive points to the interest of the CTT (Post, Telegraph, and Telephone), and of Salazar´s meddling, in the radio clubs extant at the time. Radio clubs submitted and addressed these documents to the CTT in 1954–55. One additional report from Rádio Clube de Moçamedes, written to the Commissão Coordenadora de Radiodifusão de Angola (Coordinating Commission for Radio Broadcasting in Angola) formed in 1961, was entered later. The CTT, the governor, and Salazar required a clearer sense of what broadcasting in the territory existed. Responding to what must have been a short survey-type query, all letters included, with small variations, (1) the club's history; (2) broadcast range; (3) programming; (4) equipment; (5) financing; and (6) printed propaganda/literature. Almost all provided maps of their broadcast range and photos of their installations. Some included architectural plans, lists of equipment, and broadcast schedules. A few noted they received bits of funding from the CTT or the governor's office, but the inquiry itself signals that the state had no comprehensive knowledge of the clubs or broadcasting.[106]

Some of these texts aligned radio club participation with the national Portuguese project or expressed nationalist sentiment, but just as many evidenced a sense of local and regional rootedness. Clubs sought regional connections, advertised local businesses, and broadcast community events. The Moçamedes club, for example, had plans to broadcast in English, Afrikaans, and German to better meet the needs of audiences it reached in South-West Africa (Namibia) and South Africa.[107] Diamang (the Belgian Diamond concessionary company in Angola's east) set up its own station to broadcast classical music with a particularly "airy and landscape" sensibility evocative of the African environment. While none of its broadcasters were particularly versed in nineteenth-century Russian composers, they thought this music best suited their new environs.[108] Various sonic lines flowed across regional and international borders; ones that were not always Portuguese first. The concentration of clubs in the south oriented bands outward and around the territory.

For the period before independence, the Portuguese state published annual statistical information that included data on radio broadcasting and the number of radios registered in the territory. Statistical material for the last fifteen years (1958–73) before independence shows a growing state interest in data on radio broadcasting and slight shifts in the categories of information collected. Bureaucrats used radio registration as the basis for counting receptors. It is quite likely that actual numbers exceeded those recorded and that radios typically served more than one listener, whether or not people listened collectively or serially. Still, radio density was very low. In 1972, the state registered 87,410 radios, more than half in Luanda, for a total population of around six million.[109] The state collected information on the number of broadcasters and employees per broadcaster, minutes or hours of broadcast time in general, and of particular types of programming (music, news, theater, and talk).

In the last years of Portuguese rule, radio ownership grew steadily (with a drop in 1966 that recovered by 1968), and broadcasting hours increased too. The broadcasting terrain, however, had largely been established by the late 1950s, with one exception. In 1964, one new station appeared in Huíla Province. This was Rádio Comercial. It meant that Huíla had two broadcasters, whereas the rest of the country's regions each had one, except for Luanda's three (one radio club; the Official Angolan Broadcaster, EOA; and Rádio Ecclesia).

Between 1961 and 1962, the statistical categories changed. Before 1961, the state counted all broadcasting posts (this likely included military and commercial establishment transmitters that would have used radio for internal communications), but in 1962 they began to list only broadcasting stations. Reflecting the new concern with the war, this may have been a way of more precisely tracking who was broadcasting news to general audiences, leaving internal communications in private businesses and in the military to a different kind of oversight.

The state tried to catch up with the dynamic radio scene that had developed during the 1960s in Sá de Bandeira and throughout the territory. Alongside the Huíla Radio Club, two entrepreneurs from Sá de Bandeira opened Rádio Comercial in the 1960s with the ambition of creating an Angola-wide, publicity-driven station. This was the first fully commercial station in the territory. They opened with a bang: a 10 kW shortwave transmitter, a 1 kW medium-wave transmitter, and 10 kW FM transmitter.[110] A telegram dated 3 November 1964 announced the new broadcaster, officially greeted Salazar, and offered "modest professional collaboration from this parcel of Portugal of whose future we're confident."[111] This was a vote of participation in the national struggle against the "enemy" anticolonial liberation forces.[112]

In the end, Sá de Bandeira was too isolated, too narrow, and too conservative to generate the territory-wide coverage and the profits Rádio Comercial needed to thrive. As a result, the broadcaster relocated to Luanda and kept a delegation in Sá de Bandeira.[113] Pioneering a new format, they rented airtime to producers. Radio enthusiasts such as Emídio Rangel in Sá de Bandeira, with his Estúdio A, and Sebastião Coelho, with his Estúdio Norte (originally from Nova Lisboa/Huambo but living in Luanda), stepped in to produce programming. According to Fernando Alves, Rádio Comercial did not have a clear editorial line. They wanted to be hip, forward-looking, and fast-moving. Alves produced a show, *Nocturno 73*, for Rangel that aired on Rádio Comercial. This nimble form of radio production interested young producers because it offered them space to do something different.[114]

The format worked. Rangel's program *Nocturno 73* had seven employees, as many as all of Rádio Comercial. Such growth fed innovation and professionalization. Rangel sent Fernando Esteves to the 1972 Munich Olympics from Angola via Portugal, where he waited for two days for a journalist's credential and never received it.[115] Esteves went anyway. By the time he arrived in Munich, they had already arranged all the financial support he needed. No other Angolan radio, including the Official Broadcaster (EOA), sent anyone or even imagined doing so. Rangel had vision. He could make things happen. In Luanda, Sebastião Coelho played this role.

Coelho got his start in Huambo. In 1951, Radio Club Moçamedes hired him for two positions: head of production and principal announcer. He was the first Angola-born professional to hold a senior position. Radio Club Huambo hired a radio professional from Lisbon.[116] But Portuguese professionals brought more trouble than they were worth. Eventually Angolan radio broadcasters, with the expertise they developed in the thriving scene, were scouted for jobs in Portugal.[117] Coelho returned to Huambo in 1960, another move signaling the dynamism between clubs. That year he started broadcasting *Cruzeiro do Sul* (Southern Cross), "uma nova rádio para os angolanos" (a new radio for Angolans). This was the first radio program directed to African listeners; broadcast in an African language, Umbundu; and employing black African radio announcers: Judite Luvumba and José Castro.[118] The local beer company CUCA sponsored the program.[119] Coelho played Angolan music, recording musicians and playing their music on-air. Luvumba and Castro performed folktales and oral traditions. Stories from Huambo's neighborhoods, local football matches, and the problems of urban life animated the airwaves. Skeptical of his motives, the PIDE arrested Coelho, but the show continued on the air.

After his release from prison in Luanda, the PIDE forbade Coelho from returning to Huambo. Coelho turned fixed residency in Luanda into an opportunity. Again taking recourse to CUCA, he secured funding for another show of Angolan music, this time broadcast from Radio Club Luanda,[120] *Tondoya Mukina o Kizomba* (There's a Party in Our House Tonight), in coordination with Norberto de Castro, João Canedo, and Carlitos Vieira Dias. The show's title in Kimbundu nodded to the regional African language and the language predominant in Luanda's *semba* (urban popular Angolan music).[121] With his other program, *Café da Noite* (Night Café), "boa música em boa companhia" (good music in good company), Coelho and his collaborators amassed a set of recordings that would serve as the basis for a recording business.[122]

It is no coincidence that several of the best-known names in late colonial Angolan radio emerged from the cities of the central-southern part of the territory. Alves, Rangel, and Coelho (and Leonel Cosme, David Borges, Diamantino Pereira, and Norberto de Castro, among others) came from Benguela, Sá de Bandeira, and Huambo, respectively. The distance of these places from one another and from the colonial capital at Luanda fueled the desire and need for communication. Sá de Bandeira, with the largest white settler population by proportion, was likewise the town with the most vibrant radio scene. Distance and insularity, a world both local and stranded, and access to technology motivated these young men to not only tune in but to broadcast.

FAST CARS AND WHITENESS

Radio clubs broadcast local automobile races. These offered a source of local pride and connected settlers in cities across the territory. Initially a phenomenon of the urban bourgeoisie of the period, the activity extended to the larger white community: well-known doctors owned cars, cities had their own renowned drivers, and factories their own teams. The Emissora Oficial de Angola (EOA; Official Angolan Broadcaster) broadcast all of the races, and radio clubs covered local races with four or five journalists per race.[123] In 1973 two racetracks opened: one in Luanda, another in Benguela.[124] Built by a Brazilian architect who also designed the track in Estoril, Portugal (it opened months after the two in Angola), the racetracks expressed the youthfulness and the cosmopolitanism of these cities.[125] According to Marcelo Bittencourt and Victor Melo, they marked the zenith of car racing in Angola. The racetracks united economic, political, and cultural interests at the height of petroleum's growth in the late colonial economy.[126]

Portuguese settlers pulled in international forms of white modernity that offered alternative identities to the Catholicism and conservatism of Salazar's (and then Caetano's) fascist regime and life in Portugal. If Africans who lived under the repressions of a racist and fascist colonial system looked beyond the metropole for cultural models, it is not surprising that whites who identified as Portuguese or Angolan did the same.[127] Car culture was international: fast, flashy, predominantly anglo- and francophone. Lolas, BMWs, Fords, and Porsches conquered space with speed and stretched cosmopolitan forms of whiteness to the Angolan territory.

Bittencourt and Melo argue that motorsports exemplified the links between the development of a distinctly Angolan white bourgeois culture and a separatist politics. Over the course of the late colonial period, colonial state and settler interests conflicted as much as they overlapped. Bittencourt and Melo trace this in the press, legal decrees, and publications associated with auto racing (and car touring more generally). Motorsports sprouted from the matrixes of their Portuguese structures, but by the late colonial period, their Angolan variant outpaced and outshone the sports in the metropole. The war drove greater investment in the territory, generating economic growth, white immigration, and fueling white bourgeois culture. For the local white bourgeoisie, motorsports offered a way to knit together distant areas of settlement, to know the whole territory, and to promote local resources and businesses.[128] Drivers came from Portugal to compete in Angola where the climate allowed year-round driving and the promotional market on radio was highly developed.[129] Publications distinguished between Portuguese drivers and Angolan drivers (until 1974 all of them white).[130] Like the radio clubs, they evinced identification with the Angolan territory and emotional and material investment in it.

Auto clubs and radio clubs fed this phenomenon. What in both instances commenced as hobbies—drivers and radio broadcasters supporting their own practice to fill leisure time—began to move in the direction of a more professionalized and commercialized system in the late colonial period. As Bittencourt and Melo point out, by the early 1970s, petroleum companies competed to provide gasoline for races, cars boasted advertising for Angolan businesses, and local and multinational companies sponsored events.[131] Similar shifts occurred in radio and in music.[132] With this cultural and economic dynamism, it is not surprising that a local white politics, which could see a future distinct from Portugal and tied to Angola, emerged. Bittencourt and Melo conclude:

> Motorsport was fundamentally an expression of the structuring of a white elite that, even though it was not less racist or necessarily concerned with integrating blacks and mestiços, developed a discourse about proximity to this place. Involved with the economic expansion of the province, it advocated greater respect for local peculiarities, to a greater or lesser degree focusing on provincial autonomy, but at the end of the day, it was overcome by the process of independence.[133]

Given that Salazar's Estado Novo excluded settlers from governing the colony (unlike the other settler colonies in the region), that it discriminated against those born in the territory, and that they built a locally rooted but outward-facing world for themselves, settlers sought other political solutions. During World War II, some settlers wanted to separate from Lisbon's political rule and politically approximate South Africa, according to Pimenta.[134] The Angolan United Front (FUA), founded in Benguela in 1961, drew on the left and from the local opposition to Salazar (in the 1958 election, Benguela was the sole district to vote for opposition candidate Humberto Delgado).[135] FUA advocated a period of autonomy in transition to full independence. Allied with the MPLA, FUA worked clandestinely against Portuguese colonial rule from both within and outside Angola (in Belgium and France among white Angolan students and eventually in Algeria).[136] Their publication *Kovaso* took the Umbundu (the dominant language in the central highlands region) word for "forward" as its title.[137] But FUA's existence was short-lived. It fell apart after the MPLA's National Conference in 1962 when Agostinho Neto assumed leadership.[138] MPLA's political dominance and official history have muffled the sounds of those cars, white folks playing Angolan music, and neocolonial imaginings of independence.

THE SOUND OF THE LATE COLONIAL STATE

White settlers connected to one another, to other settler communities across borders in the region, and cast a strong voice on tenuous waves to the home country. Jingoistic call signs and patriotic gestures could not disguise that many practitioners nurtured transnational links, defined an identity in local practices distinct from metropolitan Portugal, and that some even identified as Angolan.[139] Similar to the creoles Benedict Anderson discussed, they "imagined community" in the Angolan territory and in the region. When broadcasters in Moçamedes petitioned the Post, Telegraph, and Telephone (CTT) office to broadcast in Afrikaans, they hailed communities in South and South-West Africa. This complicates any facile sense of Portuguese empire or nation. It sounds out other vectors of identity and community-making, different sorts of political affinities, and an expansive sense of whiteness in formation in the Angolan territory and the region.[140]

The "tensions of empire" crackled along the fissures of interest and identity, to be sure, and they radiated across the territory around the question of broadcasting.[141] The Portuguese colonial state needed to convince Portuguese settlers in the Angolan territory, at least some of them, as much as African inhabitants, that it acted in their best interest. The colonial state could not take settler support for granted.

The Portuguese colonial state treated radio in Angola as an afterthought. Despite the entreaties of Henrique Galvão (and after him António Ferro), the erstwhile director of the EN (National Broadcaster) and champion of the colonies, and the requests of settlers, Salazar was not interested in broadcasting to or in the colonies.[142] Henrique Galvão, a man with rich colonial experience, tried several times to expand broadcasting to the colonial territories, but to no avail. Caro Proença, the last technical director of the EOA (Official Angolan Broadcaster), and his wife, Sara Chaves, a beloved singer in the late colonial period, recalled that local desire led to the opening of the EOA.[143] Until then, the Luanda-based Rádio Clube de Angola rebroadcast official news from Portugal's EN.

According to Nelson Ribeiro, the broadcasting fare included propaganda, a daily news bulletin, music produced by the station's bands, and fados. In 1937 the EN added *Meia Hora de Saudade* (A Half Hour of Saudade). This program recorded messages from relatives of those living in the colonies.[144] Tremendously popular, it was difficult to hear. Despite interest in the show, Ribeiro notes that international broadcasters (German, Italian, French, British) had better sound quality and greater listenership.[145] Operating with only a 5 kW transmitter (later upgraded to 10), weak infrastructure undermined popular enthusiasm for programming in the 1930s and 1940s. The EN received a 10 kW transmitter for international broadcasts in 1947. At that point most other European international broadcasters used multiple transmitters, some with 100 kW of power.[146] Put differently, the Portuguese frequency lacked inspiration and energy.

Official radio in Angola, the EOA, began to broadcast in response to the danger of growing nationalism and the Portuguese opposition.[147] This was late by European colonial standards, since World War II had set the agenda for broadcasting across much of the rest of the continent.[148] Experimental broadcasts started in 1951–52.[149] Local radio club broadcasters started transmitting from the veranda of the CTT building in Luanda's downtown. They tucked away the equipment in the CTT's bathroom and broadcast only at night and on weekends to avoid disrupting the CTT's regular work.[150] Programming consisted principally of radio theater and *Serão para os Trabalhadores* (Nightime Entertainment for Workers). Fados, marches, and Angolan folklore, mixed in with jokes about the natural disposition of the Portuguese driver for anticommunism (they all hate red lights), braided a variety of settler and state desires to produce a sound that was distinctly of this settler state radio, even as it hit the perfect pitch of lusotropicalism for state sponsors.[151] The EOA opened in 1953 in a house located on the city's edge at kilometer 7 (in the CTT neighborhood), close to the transmitter, if farther from the city center.[152]

This push from within the territory was root and branch of the same sentiment that produced amateur radio and aviation clubs: to combat the feeling of isolation (from other settlers and from Portugal). Pressing on the CTT, these radio enthusiasts attempted to create a local, official voice, as well as a way to be heard in Portugal.[153] While radio clubs reached out to settler communities in the region, some settlers craved easier two-way communication with Portugal and the sense of simultaneity that live broadcasting provided.[154]

Investment and organization continued to be slow in coming. It was only with the outbreak of the anticolonial wars that the regime came to understand the necessity of spreading its message. As Nelson Ribeiro put it: "Radio was then perceived as an important weapon for promoting the regime's ideology and counter-attacking the propaganda aired by Radio Moscow, and by stations operated by independence movements and political organisations opposing Salazar on the home front."[155] The sound of guerrilla stations hissed and sizzled across the borders, piquing the interest of listeners inside the Angola territory, troubling the bounds of sonic whiteness, and unnerving the colonial state. This is where chapter 2 picks up.

2 Guerrilla Broadcasters and the Unnerved Colonial State in Angola, 1961–74

GOMES'S FILM Tabu (2012), introduced in the previous chapter, captures the entangled, messy, and sensuous dimensions of the colonial/postcolonial interface. Visually arresting, it is also sonically clever. The film's title evokes another film with the same name: *Tabu* (1931) by F. W. Murnau. Murnau shot *Tabu* in Tahiti with local actors. A silent film and a box-office failure, it focused on the perils of colonialism in the Pacific Islands from the point of view of its victims. Gomes's *Tabu* stretches the chronology to the present and he reverses the section titles, situating the gaze in the perspective of the colonizer: "Paradise Lost" and then "Paradise," the first part unrolling in present-day Lisbon and the second in a Portuguese-colonized African territory (though Gomes does not tell us which one).

"Paradise" recounts the story, in flashback, of an illicit romance between Aurora, married to a colonial plantation owner, and her lover, Gianluca, a colonial mercenary. Where Murnau's film was silent for technical and technological reasons, Gomes uses sound and silence to make a point. As I noted

This chapter appeared as "Guerrilla Broadcasters and the Unnerved Colonial State in Angola (1961–74)," in the *Journal of African History* 59, no. 2 (July 2018): 241–61. Many thanks to Gregory Mann and to David Morton for their constructive feedback on the piece.

in chapter 1, both Europeans and Africans are silent in "Paradise." Viewers see Europeans talking, even if they cannot hear them. Africans are rendered mute, an apt depiction of colonialism's effects and a projection of colonial fantasy. The silence of Africans persists until the last moments of the film.

Aurora, eight months pregnant, and Gianluca flee. Outside the comforts of civilization, in an abandoned building in the bush, Gianluca's friend Mário (a close friend of Aurora's husband who has repeatedly tried to persuade Gianluca to desist) corners them. Aurora shoots and kills Mário. Two young boys lead Aurora and Gianluca to a nearby community. Amid the rondavels of a stereotypical African village, Aurora "goes mute, only screaming with the pain of early contractions," thus foreshadowing the shift in narrative voice from the colonial to the postcolonial. She gives birth, and her husband (sent for by Gianluca) appears and takes her away. Gianluca remains.

The next day Gianluca describes a radio broadcast he hears in the village: "A mysterious communiqué about Mário's death was broadcast, today considered to have been the crucial trigger for the colonial war." Then, for the first time as diegetic dialogue, we hear the stentorian tones of a guerrilla broadcaster proclaiming Mário's death as a victory for the liberation struggle: "The freedom movement takes responsibility for the death of this member of the colonialist militia as an act of war." We hear the new narrative emerging as Aurora goes mute and then, in this broadcast, Gianluca's narrative voice is overwritten by the force of an African liberation movement. The official anticolonial voice perforates the colonial sound barrier, the colonial deafness, banishing African muteness. But the "African voice" is also a representation. In the first instance, it is propaganda: the voice of the struggle, the imagined voice of the nation speaking its truth to the colonial state and settler power. And in this case, that truth is a lie. Aurora shot Mário.

This scene marks the symbolic birth of African nationalism. Its enunciation in the official, stilted voice of the guerrilla movement tears the sonically closed colonial world that dominates the second part of the film (explored in chapter 1). Yet, the choppy, stern, "speaking as if reading" voice of the guerrilla announcer reminds viewers that liberation movement voices, though they claim authenticity as the voice of the people, are also representations. We've just seen Aurora shoot Mário. If seeing is believing, hearing, in this scene, is deceiving. It tells a true lie, a strategic fiction. It poses the problem of reading colonial and official sources to tell histories of ordinary Africans.

This scene, this eruption of the voice and counternarrative of a liberation struggle, helps me introduce this chapter on guerrilla broadcasting. Chapter 2 explores how the MPLA's *Angola Combatente* (AC; Fighting Angola) and the FNLA's *Voz de Angola Livre* (VAL; Voice of Free Angola) broadcasts allowed these movements to maintain a sonic presence in the Angolan territory from exile and to engage in a war of the airwaves with the Portuguese colonial state with whom they were fighting a ground war. First and foremost, it analyzes the effects of these rebel broadcasts on listeners, be they state or nonstate actors.

A reading of state secret police and military archives exposes the nervousness and weakness of the colonial state even as it was winning the war. Portugal was fighting a war on three fronts: Angola, Mozambique, and Guinea-Bissau. Guinea-Bissau and Mozambique presented military challenges to the Portuguese military. Angola, with three national liberation movements often fighting one another, proved an easier front.[1] So why did rebel broadcasting make the Portuguese so nervous if they had the military situation under control? How did radio level the playing field? I argue that by understanding the specificity of radio as a technology—that is, the intimacy of sound and the material effects of the medium understood through the process of transduction—we gain insight into this nervousness and into the power of radio. This story picks up where Gomes's film ends.

READING FOR SOUND IN THE ARCHIVE

Gomes's film exposes the irony that the liberation movement counternarrative, claiming an authentic Angolan voice, intrudes in the colonial mise-en-scène. This is one way to understand the historian's bind when reading colonial archival sources. Writing a history of liberation movement radio requires plodding carefully in such terrain. *Angola Combatente* and *Voz Livre de Angola* have left few audio and documentary archival traces. The largest archives of radio programs are in Lisbon: in the Portuguese National Archive (Torre do Tombo) that holds the archives of the former secret police (PIDE) and in the Arquivo Histórico Militar (Military History Archive).

In 1974 Voice of America and USIA researcher James M. Kushner, in an article on African liberation radios, worried that studies ignored *how* liberation movements communicated their messages.[2] He found work of the period too focused on print. Much has changed in the intervening four decades. African radio studies thrive. Historical work on the radio is growing. The recent spate of historical work on African liberation radio addresses sources, propaganda, and listenership.[3] This new work underscores the difficulty of locating and accessing reliable archival material. Scholars point to the particularity of radio as a medium and the state of the archives as complicating factors.[4] Radio's ephemerality, the invisibility of sound, and its fleeting intimacies, along with the generally dismal state of liberation movements' sound and documentary archives, often hinder research on and off the African continent.[5]

The Angolan armed liberation movements have not archived recorded or documentary material related to their radio programs. The pressures on returning exiled movements jostling for hegemony in the period of transition between the military coup in Portugal on 25 April 1974 and Angola's independence on 11 November 1975, muddied further by foreign intervention, relegated the urgency of paper-keeping and put a premium on recycling audiotape. Liberation movement sound reels and the few written transcripts that existed were lost, damaged, reused, or simply forgotten.[6] This was true even for the triumphant MPLA.

Under such circumstances, discerning the basics of what and how the MPLA and the FNLA broadcast is difficult. Accessing listenership for any once clandestine radio is doubly vexed. Sekibakiba Lekgoathi, writing about audience and the African National Congress's (ANC) Radio Freedom, noted: "It was unlawful to listen to [the ANC's broadcasts] within the country and logically it would have been illegal if not downright impossible to conduct research on the phenomenon."[7] Systematic studies do not exist. The situation in the Portuguese colonies was no different.

Yet a substantial archive of broadcast transcripts and memos regarding listening in Angola remains. The colonial state listened in and classified the banned liberation movement broadcasts as dangerous, "anti-Portuguese," and "enemy" propaganda. Listening denoted support or, at the very least, inspired suspicion. It was illegal. But the PIDE listened diligently. PIDE agents recorded and transcribed broadcasts. Later the military and the SCCIA took over this labor. Thousands of pages of transcribed radio broadcasts of *Angola Combatente*, and to a lesser extent *Voz Livre de Angola*, sit in the PIDE files of the Portuguese National Archives and the Military History Archive in Lisbon. These are not scientific studies. What we know of the broadcasts, and of their reception, comes filtered through the interests and concerns of police chiefs, surveillance officers, soldiers, and informers.

These state agents first recorded broadcasts and then transcribed them. The PIDE destroyed the original recordings and many transcripts before the fall of the Portuguese regime in April 1974.[8] As a result, none of the files for the years 1969 to 1973 remain. Transcripts collected before 1969, namely, those produced between 1966 (when consistent transcription began) and 1968, are still available, as are those from 1974 in the months preceding the coup.[9] The PIDE and the military listened in on not only *Angola Combatente* and *Voz Livre de Angola*, but also to any other stations broadcasting material with an "anti-Portuguese character" (1966–68) and what they later described as "enemy broadcasts" (1973–74). This included the liberation movement radio broadcasters as well as Radio Moscow, the Voice of Nigeria, Radio Hanoi, and sometimes the BBC.

The Torre do Tombo and military archives in Lisbon hold the transcripts but no recordings. With the sound gone, how, then, do we read these documents? Transcripts written in all capital letters tell us more about the desire to foster legibility than about inflection. The voice is lost. While transcripts tell us something about the broadcast content and transmission quality, they tell us even more about the preoccupations of their transcribers. Accompanied by commentaries about the conditions of transmission, the perceived danger of the content, and/or the range of the broadcast, the documents capture the state's concern. The SCCIA transcripts included grids and maps. Agents managed their nervousness by subjecting broadcasts to visual dissection. Grids divided the broadcasts by theme, substantive content, target audience, and additional commentary, quantifying conclusions. SCCIA agents did not transcribe broadcasts in African languages. Ignoring disturbing content in foreign tongues, these reports betray their nervousness, suppressing sounds they cannot decipher.[10] Maps of radio reception plot the targets of sound waves. We need to learn to listen to these documents, to hear what Nancy Rose Hunt calls their "acoustic register," as much as parse their language for the subjects of broadcasts and harvest them for data.[11] This is true when we read any archival document, but it is especially urgent when analyzing the effects and work of radio. The words in these reports have a tone and timbre, though not of the broadcast voice (much of which disappeared with the recordings). Instead, this tone and timbre quivers and shouts in response to radio, registering its effects.

I follow Mhoze Chikowero, who mines the colonial radio archive to destabilize the solidity of the state's story about its own project.[12] He locates the "nervous condition of colonial authority" in concerns about broadcasting and how Africans read the press.[13] Nancy Rose Hunt moves in a slightly different direction when she theorizes the Belgian colonial state in the Congo as "a nervous state." A nervous state of being "taut, a nervous wreck"; that is, on edge, on the verge, and unstable. Hunt finds evidence of this in the state-run institutions that react to therapeutic insurgencies, on the one hand, and the enumerating, modernist visibilizing biopolitical practices of the state in health clinics that promoted pronatalist policies to correct colonial destruction, on the other.[14] In particular, she scours the archives for sonic traces of violence to loosen the grip of the visual on how journalists, scholars, and humanitarians framed and imagined the Congo Free State and Belgian

Congo. In the space that sound opens, Hunt discerns the effects of a nervous state; laughter destabilizes its routine violence and exposes its affect.

Attending to sound in the PIDE and military archives of *AC* and *VAL* broadcasts means thinking at the interface of sound, nervousness, and nationalist insurgency, both real and imagined. Despite the fact that the Portuguese were winning the ground war—a narrative repeated in news reports in the Portuguese press—the archives disclose Portuguese military and secret police insecurity and nervousness. Guerrilla radio, I argue, produced this.

ANTICOLONIAL WAR

Chapter 1 presented a history of radio's origins and its relationship to white settler life. This chapter studies how the MPLA's and FNLA's broadcasted counternarratives affected the colonial state, namely, the military and secret police. Portugal under the Salazar dictatorship was a gray, inhibited, and repressed world, the opposite of white settler life in the colonies. In Angola, and other Portuguese colonies, political repression and economic exploitation permeated colonial society, especially the lives of Africans. The police suppressed nationalist organizers in 1959 in the Processo de 50 (Trial of 50). The state arrested and sent some nationalists to Luanda prisons and others to Tarrafal Prison in Cape Verde. The PIDE arrived in Angola shortly thereafter,[15] but it did not stop insurgency.

In early January 1961, people in Malanje revolted against colonial administration and forced cotton production. They attacked colonial state agents (*cipaios* and translators), destroyed identity cards and cottonseeds, and damaged stores and the concession company's property.[16] They refused to pay the head tax or obey the local administrator.[17] According to colonial sources, people in the area mobilized around a prophetic movement that announced the arrival of Maria, a woman who would deliver freedom from Portuguese oppression. These sources recount that António Mariano, originally from Malanje but recently returned from the newly independent Congo, galvanized a following preaching Maria's return in what some scholars refer to as "Maria's War."[18] Historian Aida Freudenthal called this an "anti-colonial revolt permeated by an ethno-nationalist ferment," though it lacked a nationalist or even a clear political program.[19] Margarida Paredes is the first scholar to have interviewed participants in the events. One woman, when asked, "What was 'Maria's War'?" responded that it "was a code brought by the Congolese to run off the whites."[20] Whether a prophetic movement or a code name for a call to action, the revolt evidenced, if not prefigured, a robust, cross-border communications network—one that straddled the material and the spiritual worlds—well before the advent of guerrilla broadcasting.[21]

In a revisionist reading of these events, Aharon de Grassi sees the revolt as part of a nationalist mobilization that connected Luanda, small and large towns in Malanje, and Congo-Kinshasa, through migration, activism, and contract labor.[22] Malanje was a "crossroads of nationalism."[23] The mobilization and revolt surfaced territorial, not just local, processes.[24] Protesters said they had had enough with five hundred years of Portuguese rule and would take orders only from Joseph Kasavubu (president of the newly independent Congo) through his messenger Maria.[25] De Grassi argues that this was, in fact, the start of nationalist revolt and that it spread throughout the region as refugees from the area moved out, sometimes across the border to the Congo, and created a base for the Union of Angolan People (UPA, the predecessor of the FNLA).[26] African residents of the Angolan territory pursued and nurtured ties of language, cultural practice, and commerce across the colony and region. If the colonial state and Portuguese settlers had been largely deaf to African discontent, this clamor finally perforated that silence.[27]

In the face of widespread revolt, the Portuguese colonial state sent in the army, police, and air force to bomb villages in the Baixa do Kassanje between February and March. Estimates of the dead ranged between the hundreds and thirty thousand.[28] Freudenthal elaborates, suggesting that thousands were killed and many thousands more fled to the Congo, seeking shelter from the bombings and continued Portuguese repression.[29]

In the midst of this, though disconnected from it, on 4 February 1961, a group of activists attacked two prisons in Luanda, aiming to free some of the political prisoners jailed in the Trial of 50. Organized in part by the Catholic canon Manuel das Neves, and in coordination with UPA, the colonial state again responded in a repressive fashion. In Luanda, Portuguese civilians received arms and meted out reckless violence on urban elites they imagined were associated with the revolt.[30] Meanwhile, in the country's north, on 15 March 1961, armed UPA men based in the Congo, and men they had mobilized locally, attacked village border posts, coffee plantations, and administrative offices and commercial establishments in a wide area that included the Dembos, São Salvador, and Carmona.[31] David Birmingham described the attacks as "the largest colonial uprising to be experienced in any part of tropical Africa."[32] UPA militants massacred hundreds of Portuguese plantation owners and Ovimbundu workers. The colonial state and military responded with unprecedented violence, decapitating Africans and posting their heads on stakes to terrorize the local populations, and killing thousands in retaliation.[33] Feeling under siege, and unwilling to negotiate, the Portuguese colonial state again reacted with violence.[34] This would be war.[35] The carnage measured "ten times larger" than the Mau Mau war in Kenya, Birmingham estimates.[36]

The three attacks brought to the surface systems of coordination and communication that the Portuguese colonial state had neither glimpsed nor heard. The imprisonment of nationalist agitators after the Trial of 50 temporarily subdued anticolonial politics.[37] While the events of 1961 heightened surveillance and security, they exposed desires illegible to the Portuguese state even as their police scoured the land for signs of communist infiltration. They might have had more success had they listened instead of just looked. Tuning in to foreign broadcasters and producing music that in its language and sounds defined what it meant to be Angolan, urban residents had already begun to imagine themselves as a distinct group with specific political interests.

Arriving in the territory in 1959 after the Trial of 50, the PIDE followed radio broadcasts in the region. Initially concerned with the Portuguese opposition, when *Angola Combatente* began broadcasting in 1964, PIDE agents tuned in. While *Voz de Angola Livre* began broadcasting consistently in 1965, the PIDE did not begin systematically recording and transcribing its broadcasts until 1966.[38]

GUERRILLA RADIOS AND THE WAR

The events of 1961 forced the nationalist movements to consolidate themselves in exile. Based outside Angola's borders, they needed a way to communicate with militants in the territory and to spread their message. The external services of newly independent countries offered airtime to nationalist movements fighting the white settler redoubts of southern Africa, giving them access to international airwaves.[39] Intermittent broadcasts came from Ghana. Consistent transmission occurred only after the MPLA settled in Brazzaville, and the FNLA, already based in Leopoldville/Kinshasa, established a relationship with the broadcaster.

Angola Combatente's Program A began broadcasting from Brazzaville in 1964.[40] Program B broadcast from Dar es Salaam beginning in 1968, but with poor sound quality.[41] Program C, broadcast from the MPLA base camp Vitória É Certa (Victory is certain) in Lusaka, started in 1972 to cover the eastern areas of Angola in that region's languages (Luvalu, Tchokwe, Umbundu, and Portuguese).[42] An outgrowth of the movement's Department of Information and Propaganda, AC was the bailiwick of party intellectuals (in Brazzaville, Aníbal de Melo and Deolinda Rodrigues, later Adolfo Maria; in Lusaka, Paulo Jorge, Mbeto Traça, and Ilda Carreira).[43] None had previous radio experience. Some had worked in journalism and others were committed militants with more education than average cadres, so the movement employed them in writing broadcasts, newsletters, pamphlets, and news releases as well as in producing photos and film clips. Propaganda work often took place in coordination with foreign journalists.[44]

The MPLA broadcast Program A from the radio network "Voice of the Congolese Revolution in Brazzaville." This radio network had the strongest transmitter on the continent, a well-used and guarded inheritance from the World War II–era Free France movement. The independent government of the Republic of the Congo had maintained and expanded the station's technical capacity. Like other newly sovereign African states, the Republic of the Congo supported the liberation movements of those still under colonial rule and therefore put its broadcast power at the disposal of the MPLA.[45] A PIDE agent transcribed an AC broadcast from Brazzaville in which the tagline underscored that solidarity.[46] The PIDE agent closed by reporting sound interference, likely atmospheric. Reliant on the ionosphere to bounce electromagnetic waves to faraway places, what shortwave gains in distance it often loses in quality. But as the many files of transcripts attest, plenty of broadcasts came through loud and clear.

The FNLA broadcast *Voz de Angola Livre* from the Voice of Zaire radio stations in Kinshasa and Lubumbashi. The *VAL* broadcast in the name of the Revolutionary Government of Angola in Exile (GRAE), formed by the FNLA in 1962. According to Holden Roberto, former FNLA president, party militants broadcast from Leopoldville/Kinshasa between 1961 and 1974. PIDE transcripts of the program date to 1966, though the documentary film *Independência* (Independence, 2015) includes a broadcast from 1963, and a 1964 PIDE report from Malanje mentions the broadcasts.[47]

Two men ran the *VAL* program.[48] Both received training in radio at Zaire's National Radio Station. *VAL* broadcasts, like those of AC, pushed back against Portuguese propaganda, addressed Portuguese soldiers and encouraged them to desert, followed the EOA closely, sent news of battlefront gains and losses, reported on the movement leaders' diplomatic travels, and communicated messages to Angolans within the country.[49] Much more of *VAL*'s programming than that of AC addressed the sizable Angolan population in exile in Congo/Zaire and announced or reported on local community events. In what would later take on an ironic ring, one *VAL* broadcast attacked American imperialism.[50]

The PIDE and military archives contain more *Angola Combatente* than *Voz de Angola Livre* program transcripts. The MPLA broadcast more regularly. PIDE and military documents highlight the MPLA's investment in propaganda, particularly via the radio. For example, a military report on psychosocial action from late October 1968 mentioned that "among the EN [enemy] psychological activities for the period, the MPLA continues to distinguish itself for having best undertaken them, especially in radio broadcasting."[51]

The military mounted a surveillance operation as part of its counterinsurgency strategy.[52] In the wake of the Baixa do Kassanje attack, a working group composed of military, PIDE, and civilian elements proposed the formation of the SCCIA to organize information from the various surveilling agencies. Political infighting among the secret police, military, and SCCIA hindered coordination.[53] Nonetheless, in Angola these agencies communicated with one another frequently and more regularly than in Mozambique or Guinea-Bissau.[54] Surveillance proved effective. By the late 1960s, the PIDE had dismantled urban-based MPLA-, FNLA-, and UNITA-affiliated cells.[55]

Both the PIDE and the SCCIA undertook radio surveillance. PIDE and SCCIA agents forwarded transcripts of radio broadcasts to the military, which had its own radio listening unit—the Broadcast Reconnaissance Command (CHERET). The PIDE sent individual transcripts while SCCIA forwarded monthly situation reports that combined transcripts, commentary, and, by the late 1960s, maps that showed broadcast reception and quality in the territory. This duplication, if not triplication, of labor bespeaks an inefficient need to record. The PIDE followed listening practices in the territory, while SCCIA focused on broadcast content. Meanwhile, the military also worried, in a general sense, about listenership and the impact of what they called "enemy propaganda" on soldiers, the white population, and Africans.[56]

CLANDESTINE LISTENING: WHERE ARCHIVE AND MEMORY MEET

Luanda is twice marked with the history of liberation radio. The brutalist modernist edifice of what is today Rádio Nacional de Angola (Angolan National Radio) is shown in figure 2.1. Built by the former colonial broadcaster, EOA, to counter the disturbing broadcasts of the liberation movement radio broadcasts, RNA is still at the center of the country's communications network. A block away, a modest, well-maintained representation of the MPLA (Popular Movement for the Liberation of Angola) radio program *Angola Combatente* (Fighting Angola) is part of a mural (see fig. 2.2) on the

FIGURE 2.1 Rádio Nacional de Angola (formerly Emissora Oficial de Angola). The building was under construction from 1964 to 1967. Angolan architect Fernão Lopes Simões de Carvalho designed the building. Photo from collection of Fernão Lopes Simões de Carvalho and used with his permission.

FIGURE 2.2 Mural on the wall of the Luanda military hospital. The whole mural recounts the national liberation struggle from the official MPLA narrative. This panel shows the significance of media (newspaper and radio) in that fight. Initially painted by Teresa Gama and a number of art students shortly after independence, it is today one of the few maintained works of public art from that period. Photo taken by Margarida Paredes and used with her permission.

walls surrounding the military hospital. This work of public art recounts the history of the MPLA's struggle and triumph against Portuguese colonialism. The modernist radio station is the product of late colonial counterinsurgency infrastructure and the mural the result of a postcolonial socialist popular art mobilization and official public history. The former bespeaks the colonial state's nervousness, the latter the need of the new MPLA state to project a sense of certainty about history and memory.

Listening to *Angola Combatente* remains a proud, vivid memory in official and popular discourses about the exiled liberation struggle. Commemorated in the public history mural on the wall of the military hospital in central Luanda, the image of the radio with two listeners next to a man reading the MPLA's paper *Vitória ou Morte* (Victory or Death)/Brazzaville/ Último Comunicado da Guerra (Latest War News), is a key part of the MPLA's official narrative of the national liberation struggle. While the current MPLA regime and new businesses raze other artifacts of the early days of independence (signs, marks to denote building occupations, monuments), painters refresh this mural, maintaining visual narrative certainty against street-level dust and the to-and-fro of daily life in Luanda. Radio broadcasting, as the newspaper in the mural suggests, played its part in a propaganda strategy that involved information broadly cast: pamphlets, an international network of connections with socialist political movements, international journalists, and exiled centers of study.

Three lines joined in the upper corner of the mural denote an interior listening scene, reminding viewers that people hid their listening. A 1968 memo from the Steering Committee on Radio in Angola (CORANGOLA) to the PIDE confirmed such practices. Informers encountered a rumored, duplicitous listening practice in the *musseques* (Luanda's informal neighborhoods predominantly populated by Africans): "Many radio owners have two apparatuses: one located in the entry room tuned to a Luanda broadcaster and another, located in the back of the home, tuned to Brazzaville."[57] Memory and archive intersect.

People often hid their radio listening in extreme ways (at least in the memories they recount), and while the AC and *VAL* broadcasters were out of sight (located in exile), and the exiled movements of which they were a part were largely invisible to those inside the country, the broadcasts themselves were not a secret. That would have undermined the broadcast's utility. Invisibility cloaked listening practices in order to better spread the message. Militants and interested listeners dissimulated in order to tune in.[58] Some people later met secretly to pass along what they had heard. Liberation movement sympathizers, and their opponents, created networks of communication in which guerrilla radio broadcasts constituted one node. Transmission of messages from the radio sometimes had visible effects. Historian Marcelo Bittencourt tells us that "in 1966 slogans and words of support for the MPLA started appearing posted or painted in public places," offering material evidence that AC's message sometimes came through loud and clear.[59]

Individuals recall that the guerrilla stations catalyzed their political awakening. Rodeth Gil said that the MPLA's broadcast informed people of the anticolonial war, something not reported in the colonial press.[60] Based in exile, movements broadcast news that could perforate the Portuguese state censorship woven into the already insulated broadcasts of radio clubs and state radio.[61] Lote Chivava Guilherme "Sachikwenda" recalled young boys in groups of four or five gathering around the radio to listen to the broadcasts of the movements in exile. This was how he and his peers learned about the war.[62] Radio sparked political awakening. Ruth Mendes remembered hearing about the liberation movements and tuning in as a twelve-year-old girl: "Then we started listening to *Angola Combatente* without the adults knowing."[63] The broadcasts sparked fear of reprisals; young people hid their listening from the PIDE and their elders.[64]

Scenes of secret or semisecret listening, shrouded in fear, often anchor memories of guerrilla radio. Lekgoathi found similar "listening cultures" in South Africa where listeners met under cover of night, to avoid detection by police and informants, and later debated issues from the ANC's Radio Freedom broadcasts.[65] Manuel Faria, owner of a recreational club in the musseque Sambizanga, recalled driving to a soccer field, turning off the car's engine, and tuning in on a transistor inside the car in the field's dark, empty expanse. The composer and musician Xabanu listened with his political cell; and Alberto Jaime, a civil servant, described AC as a call to action, but a potential "death certificate" for those caught listening.[66]

A somewhat uncharacteristic scene, because of the size of the radio and the inversion of its location, appears in Zeze Gamboa's 2013 film, *O Grande Kilapy* (The Great Swindle). The playboy protagonist and nationalist Joãozinho comes home late one night and startles his father, whom he finds with his ear close to the family's large, tabletop radio, listening to the AC broadcast. Jardo Muekalia, today a senior member in UNITA, offers another tale of the masculinist and paternalist construction of nationalism through radio. One night in 1968, nine-year-old Muekalia finds his father alone on the patio, listening to a broadcast at low volume. Approaching, he hears the call sign from Brazzaville. When his father realizes his son is near, he turns off the radio immediately, asks him to repeat what he heard, and swears him to silence. Noticing his father had not changed the frequency, Muekalia returns to listen:

> They continued to talk about the MPLA and colonialism. In any case, it was all Chinese to me. All I knew was that it was a dangerous broadcaster. From then on, I listened now and then to the dangerous broadcaster, accompanied by my brother Tiago, until one day my father caught me. I had never seen him so angry.[67]

Weeks later the PIDE arrested his father, a Methodist minister. PIDE agents accused Minister Muekalia, who was raising funds for the local church mission, of using the money to support the *turras*, Portuguese slang for terrorists and a general reference to the nationalist movements.

Material from the PIDE archives reiterates the trope of hidden listening, if not the sense of fear. But "semiclandestine" better describes the listening practices police and informers often encountered. In their reports, police informants disdained listeners' intrepidness. In Luanda in 1967, the author of this memo expressed exasperation at public displays of listening to programs contiguous with those of the MPLA: "Enough of the shamelessness of natives who walk around with transistors in hand in the middle of the *musseque*, listening to broadcasts from the Congos—Kinshasa and Brazza. As for the MPLA broadcasts, they are listened to inside the home. The influence among natives is great, increasing their euphoria in favor of independence."[68]

PIDE agents rarely referred so directly to independence. More typically they used terms like "insurgency," "terrorism," or "enemy broadcasts" to describe the desires of Angolans for independence. In other words, even as they transcribed broadcasts about nationalism, agents sifted that content through an ideological frame that delegitimized nationalism.

One officer reported that in the musseques, the PIDE commonly "surprise whole families grouped around the radio listening devotedly as if it were a religious cult."[69] PIDE officers often referred to the population of the musseques as a whole, as if they were a kind of horde, one person indistinguishable from the next. For example, six months after the report cited above, a different informant reported that "almost the whole population of the *musseques* can't stop listening to the Brazza broadcasts."[70] Lumping together musseque residents and all programming from the station, the informant equates listening with "insurgency." In Sambizanga, "in various backyards, it was noted that avid subversives were found, listening to Brazza, and in one, there were five individuals with a radio inside a basket, listening to the broadcast in religious silence."[71] Some committed political activists cultivated a sense of religious devotion around their listening.[72] Whether this was true for the average musseque listener or was the projection of those recording the act is harder to know. But people did gather at a regular time, often in different places, to listen quietly and with focus. The association between radio broadcasting, listening, and religion or spirituality is a tight one. Aside from the widespread use of radio by religious organizations worldwide, the communications scholar John Durham Peters notes the mutual exchange between early radio enthusiasts and spiritualists in the late nineteenth century.[73]

The memories of Angolans and the PIDE documents highlight the occlusion, not the invisibility, and the muffling, not the total silencing, of listening. Faria tuned in openly, but in the dark. Minister Muekalia listened outside, but at low volume. And listeners shared the contents of the broadcasts, relishing and debating details. They passed along the coded names of friends, or friends of friends or contacts, who had managed to make it to exile, or news of victories in international diplomacy. Muekalia's reminiscence points to youthful curiosity and the attraction of danger, in what sounded like a foreign language, in the MPLA's alternative interpretation of lived reality. Like Jaime, Muekalia stresses the tremendous risk faced by nationalist sympathizers or those just curious about other points of view. In the late colonial period, many people dreaded the PIDE. Transgressing the law, even when impelled by a sense of righteousness, or curiosity, still involved fear.

Indeed, listening could lead to arrest. An example from 1967 in the central Angolan village of Mungo, near the city of Nova Lisboa (a postindependence UNITA stronghold), illustrates this. Around 7 p.m. one August night, the local administrator walked into a bar to buy cigarettes. As he entered, the barman, Timoteo Chingualulo, turned down the volume on the radio and then his friend, a lab nurse from the local health delegation, António Francisco da Silva "Baião," changed the station. After the administrator exited, they tuned back in to the original station: Radio Brazzaville broadcasting the MPLA's program. The administrator could hear it from his veranda. The local police detained both men and the offending radio.[74] While the police had no evidence that the men were liberation movement members, "it is inferred that the accused are partisans of an independent Angola, who, for now, are trying to satisfy their ambition by sending out the Brazzaville broadcasts publicly."[75] Listening marked one as a subversive.

Press censorship and PIDE surveillance encircled radio listening with risk. The fear and perceived danger involved in tuning in to the guerrilla stations means that former listeners sometimes recall listening to liberation movement radio programs as a mode of participation in the struggle for liberation. They offer the memory of listening as a symbol of having been a part of or aligned with the exiled movements, particularly that of the MPLA, where service in exile continues to be a significant marker of party membership status.[76] Some people proffer listening as a badge of long-standing loyalty to the ruling party. At a book launch for a novel by an Angolan author in Lisbon, Portugal, in April 2016, one attendee exclaimed: "I am from the MPLA. We listened to *Angola Combatente*!" Here the speaker collapses party affiliation, if not militancy, and listening. If listening generated real fears, and listeners ran risks that some felt demanded recognition or recompense, PIDE officers had their own affective responses.

THE COLONIAL STATE UNNERVED

The PIDE sometimes arrested listeners, but not always. In 1964, an informant found a nurse working in Luanda rejoicing over news he'd heard on *Angola Combatente*. The informer sent a memo up the intelligence hierarchy, but it did not have immediate repercussions.[77] As the arrest of Chingualulo and Baião demonstrates, PIDE files offer evidence that the practice of listening extended beyond the politically conscious and engaged activists of urban areas. Even a cursory reading of the many thousands of pages of documents of broadcast transcripts, interrupted listening séances, decrees about how to handle the new broadcasts, and reports of rumored tuning in brought to the PIDE amplifies the trope of hidden listening and complicates our understanding of what it meant.

It is difficult to know precisely how many people listened and how widely. PIDE transcripts came from nearly all provinces, but accounts of Africans tuning in did not. In a study of broadcasting in the Portuguese colonies, Alexander F. Toogood reported low radio density in Angola:

> Considering that in 1971 there were only about 95,000 receivers in the country, one-third concentrated in the Luanda district and mostly in the hands of Europeans, it seems doubtful that broadcasts by the nationalist movements could have a widespread impact. . . . It may be a measure of Portugal's insecurity that it nevertheless tried strenuously to counter the propaganda broadcasts by increasing Radio Angola's provincial coverage and giving financial support to private stations.[78]

If memories of listening seem exaggerated relative to the actual numbers of radios in homes (even if one radio set served many ears), so, too, does Portuguese "insecurity."[79] In Toogood's account this paranoia motivated action.

Beginning in 1966, reports to the PIDE about *VAL* broadcasts described them as being "of a subversive character." PIDE officers recommended jamming the broadcast, something they were attempting to do with *AC*.[80] Often issued from the Luanda delegation, it is not clear from these documents where reception occurred, though an entire file of nearly one thousand pages of transcripts of both *AC* and *VAL* from 1966 to 1968 includes nearly a third from two posts in other areas: Dundo-Portugalia, in the far eastern Lunda region (the base of Diamang the diamond concessionary); and Lobito, a port city in central Angola.[81] In other words, the sound came from all around, and listeners across the territory tuned in.

Memos relating to radio crossed the desk of the PIDE's head of office, Jaime Oliveira. He received and signed off on nearly every report of listening. He drafted memos and circulars that resound with what Chikowero calls the "nervous condition of colonial authority."[82] In May 1966, Oliveira reported on listenership, securing the very trope that dominates in memory: "In the most complete silence and isolation, uniting all the possible and imaginable precautions, the middle class, that is, administrative functionaries and servants and others, in an incredibly large number, listen to Radio Brazzaville on Sundays (the program of the subversive broadcasts)." More prudent than elite listeners, Oliveira found these "middle-class" listeners difficult to count.[83] He continued: "These broadcasts inspire them and give them ideas and oblige us to think, to judge that these [broadcasts] are no less dangerous than the armed war." Oliveira added seven "civilized" listeners, whom he listed by name—a laughably small number, he noted, given the sizable total number of listeners, including "a large number of white sympathizers."[84]

Propaganda, the war of words, of hearts and minds, captured the attention of large swaths of colonial society. Even white settlers tuned in regularly.[85] Broadcasts appealing to soldiers to desert caught listener attention and troubled the PIDE. In a series of reports from across the territory, from Pereira d'Eça in the far south to Cuanza Norte and Luanda in the center, the PIDE registered anticipation about and chatter around interviews with deserters in the late 1960s.[86] One officer noted the effects on Portuguese soldiers: some suddenly realized that they were defending the interests of elites.[87] Two months later another officer reported that an *AC* broadcast, which had wide repercussions in Luandan public opinion, had encouraged desertion on the grounds that "the army doesn't defend the Nation but the capitalism of 'Diamang,' of Oil and a half-dozen wealthy individuals because the people of Angola live in misery."[88] Another report notes that an AC interview of African soldiers who deserted the Portuguese army was "listened to attentively," and one man was overheard repeating it nearly "word for word" to a friend.[89]

In December 1966, Oliveira issued Secret Circular 52/66 to all PIDE delegations and subdelegations from the Luanda office. This circular noted the growth in anti-Portuguese radio propaganda, especially that from the MPLA. Commenting on reception, he noted that their "programs reach the Province with the best conditions for listening, giving the sensation that we are hearing a local broadcast, and a large part of the population, even the Europeans, listens closely."[90] Oliveira went on to solicit help from local post and telegraph offices and radio clubs in jamming *Angola Combatente*. At this point, he did not know if it would be effective or even possible. Having come up short both technologically and in the propaganda game, the PIDE scrambled to take the upper hand, calling on local expertise and trying to centralize information and skills in a disconnected colonial world whose coherence lay only in violence.[91]

But the PIDE had been following foreign-based broadcasters since the late 1950s. They knew of the MPLA's and the FNLA's broadcasts well before 1966, even if they were not yet following them systematically. Attempts and requests to interfere with Brazzaville's signal were not new. Neither was the nervousness. It was an unsettling, destabilizing force that would not go away. Brimming with colonial stereotypes, a 1964 report from the PIDE subdelegation in Malanje requested that Luanda insist that Lisbon prioritize "the interference [with] Brazzaville communications." The body of the report concerned "strange facts" apparent in the "life of natives" in certain Malanje towns in the preceding month. Among them: a rumor that people should not work on Saturdays and Mondays, aimed at "damaging the economy of the Province and as some kind of protest of something that is not yet clear." The head of the PIDE delegation blamed African priests, pastors, and catechists in the area. He associated these behaviors with a state of unrest and rumors that circulated years earlier in 1961 when colonial military first arrived in the area, following the uprisings against forced cotton production and the resultant massacre by the Portuguese military in the Baixa do Kassange.[92]

The PIDE's head of delegation in Malanje stacked up evidence and fueled agitation. He described the residents of Malanje villages as "arrogant, insolent, and daring" in their interactions with whites. He said that residents had been told they should all have a guitar to play and sing "the anthem of independence." Thus the "native masses" wandered around with homemade guitars, "animat[ing] the idea of latent agitation noted and confirm[ing] the state of subversion observed all around." The PIDE head of delegation then noted that those engaged in healing and witchcraft, "who work clandestinely to achieve their ends," further fomented subversion. Finally, the last link in the chain: "In mixed commercial establishments and those which sell electrical articles, it's common to see natives, some of whom we are surveilling, buying portable radios and this is due, without a doubt, to the reception conditions of Radio Brazzaville, via which the directors of the MPLA incite the mass [*sic*] to rebellion, in all manner of conspiratorial ways."[93]

To summarize, the chief of the Malanje PIDE office reported absenteeism from work on Saturdays and Mondays, the increased appearance of artisanal guitars, healing rituals, and radio sales. He linked these visible phenomena to what he could not see: rumor, clandestine (his word) witches and healers, and the MPLA on Radio Brazzaville. He read "strange behavior" as indexical of clandestine political activity. Where there was smoke, there was fire.

This document evokes associations typically attributed to Africans interacting with new technologies, like radios or cinema or medical devices.[94] Africans, colonial administrators insist, see magic in machines (so, of course, do most first-time users anywhere), use rumor as a form of communication, and are naturally inclined to musical recreation. Here the PIDE chief incorrectly divines a single cause: the MPLA broadcast. One can almost hear the MPLA leadership tittering with delight.

RADIO TECHNOLOGY AND NERVOUSNESS

The PIDE officer evinces nervousness in the face of what he calls "strange behavior." To quell his tremulousness, he writes. In the act of writing, he knits visible symptoms with a causality he locates in the invisible voice of the MPLA broadcast from Brazzaville. The solution? Jamming. Scramble the sound, interfere with transmission, block reception, calm the nerves. But his blunt diagnosis rests on shaky evidence. The uprising in the region in 1961, not linked with any organized political movement, had some ties to Congo-Kinshasa, where the uprising's founder had worked and been touched by the Congolese politics of independence.[95] De Grassi reminds us that protesters chanted Lumumba's name.[96] No evidence linked the uprising to the MPLA.

The officer clutches at causal certainty to counter his edgy state. But what precisely is the source of the nervousness? Hunt argues that "the Belgian colonial state was born from nervousness and the Congo became a nervous state."[97] She finds its beginnings in the "tense, aggressive Free state, fierce Stanley, taut officers, wrathful inebriated concession agents, and armed sentries."[98] For Chikowero, the colonial archive on radio bespeaks the "nervous condition" of the colonial state by exposing its fear that radio would inspire nationalist insurgency that undermined its authority, a sentiment present throughout this document.[99] While the Portuguese colonial state in Angola, thickened in reiterated acts of violence strewn across the difficulties of metropolitan and territorial politics and identities, certainly was a nervous state, this officer's nervousness, and that of Jaime Oliveira, and the PIDE and military, also derived from the specificity of radio technology.

Immateriality, intimacy, and transduction characterize the radio. Sound waves travel through the ether, diminish distance, and banish time. Michele Hilmes describes the unifying ambitions of broadcasting in the early twentieth-century United States: "The basic technical qualities of radio would unite the nation physically, across geographic space, connecting remote regions with centers of civilization and culture, tying the country together over the invisible waves of ether."[100] Radio broadcasting can connect across an empire despite miles and time differences or stitch together white settlements in a far-flung colonial territory, as member-based radio clubs in Angola did. But those invisible sound waves could create intimacies of a different stripe as well. Guerrilla stations, based in exile in sovereign African states, existed beyond the jurisdiction of colonial law but within broadcast range of the colonial state and the territory it claimed. Herein lies the power of the immateriality of radio, shortwave in particular, to disrupt and to unnerve. It is unruly, does not respect borders, runs roughshod over the sense of inviolable national territory. Broadcasting anticolonial propaganda, the stations of the MPLA and FNLA put the colonial state and its police and military on edge and on the defensive.

Guerrilla broadcasters caught the state off-guard and drew its attention to listening practices. If reports of musseque dwellers owning radios abounded, and PIDE officers nervously repeated rumors that some homes had two—one in the front tuned to the state broadcaster, one in the back tuned to Brazzaville—then hidden listening inside and outside the home was rife. Plenty of Europeans and black civil servants listened quietly in their homes. The capacity of radio broadcasting to produce unity and intimacy meant that it not only conquered distance but created seclusion in intimate listening spaces where listeners imagined similar scenes in the neighbors' homes or in the homes of people in the territory's other towns and cities.[101]

PIDE and military officers found a diverse set of listeners huddled around radios, alone or in groups, attentive to the nationalist movements' news, critiques, and exhortations. One military report noted the MPLA program's "electrifying" effects.[102] The intimacy of radio, the fact of broadcasting into the private space of the home, into the ear and head of the listener, set the minds of PIDE officers reeling. It made them unsettled, jumpy, and reactive.

This had everything to do with radio technology. Brian Larkin suggests: "This dynamic, whereby the agency of the object (Latour 1993) has an independence from the intentions governing its introduction, opens up the sensate, material world of the technology itself."[103] Technologies do not determine outcomes, but how they operate has material consequences. This is key to understanding how they produce nervousness. Radios are transducers. They change sound energetically as the waves move across and through them.[104] Radios transform immaterial sound waves into material effects. Perhaps what made the PIDE officers and informers so nervous was this transformation and potential transubstantiation. Bodies close to radios, ears penetrated by that energy, could be transduced and changed. Officers nervously opined that radio listening made subversives of listeners or that listening equated to subversion. Perhaps it was the fear that this transformation was less about broadcast content and ideas than about how machines affect bodies that made officers and informants and Oliveira himself so very nervous. The technical operations of radio troubled the minds of PIDE officers and informants as they touched the bodies of listeners.

Fanon's reading of radio in Algeria is apropos. In the following passage he is discussing what happens once Algerians discover the Voice of Free Algeria. The French, keen to limit listening, outlaw the sale of radios (except to those with special vouchers). The military jams the broadcasts. It is a scenario much like the one that develops in Angola a decade later. The Voice of Free Algeria must relay the first jammed broadcast via a second broadcast, sometimes a third. Listeners rarely hear a complete show. Thus "the listener, enrolled in the battle of the waves, had to figure out the tactics of the enemy, and in *an almost physical way* circumvent the strategy of the adversary" (emphasis mine).[105] Listeners captured only a fragmentary broadcast and had to create information to fill in the gaps.

Much like those who listen under desks or in dark soccer fields, then whisper to friends the next day, the Algerian listeners Fanon describes are not dead-end receivers. They are transmitters too. They are transformed into radios, John Mowitt argues when he reads this passage in Fanon.[106] Radio is not just an object; it is, as Fanon insists, a technique.[107] Radio in this sense can be revolutionary. It transforms the person. Its work could take place in plain sight, but still be devastating, still clandestine for being illegal and largely imperceptible.

South African poet Ingrid de Kok's poem "The Sound Engineer" pushes the reader into a deeper understanding of the ear, sound, and nervousness that result from transduction.[108] De Kok's poem is chilling. Choppy lines communicate how the work of listening and editing sound for broadcast reenacts violence:

> Listen, cut; comma, cut.
> Bind grammar to horror,
> blood heating the earphones,
> beating the airwaves' wings.[109]

Banal punctuation and routine acts of editing translate torture. However, the sound engineer is not the perpetrator but rather an aural witness, sonically accompanying the acts recounted. Until the body succumbs: "The sound engineer hears / his own tympanic membrane tear."[110] This is secondary, or vicarious, trauma in its psychological and physical manifestations.[111] De Kok reports this secondhand in the poem's epigraph: "Of all the professionals engaged in Truth Commission reporting, the highest turn-over was apparently among reporters editing sound for radio."[112] She expresses that which cannot be captured and contained by technologies of sound: what slips away between machine, emotion, and witnessing and is so vital to the scene.

For the sound engineer, the outcome of aural witnessing, of vicarious trauma, is physical trauma: the tearing of the tympanic membrane. The tympanic membrane receives and perceives the vibrations in the airwaves that constitute sound. Sound enters the ear canal. It touches and percusses the membrane behind which sits the cochlear nerve. Sound strikes the body's nervous system.

If the state misapprehended the effects of radio in towns in Malanje or overestimated its capacity for creating subversive activities, even as it intuited something true about how it operates, MPLA loyals in the present miss something about the radio too. The persistent trope of clandestine listening does not fully capture radio's power. For the MPLA, it offers a history of presence in the territory that never really materialized. It sounded like they were around, but were they? For those politically conscious but not necessarily prepared to run off to the front and join one of the movements, listening in was sometimes the only way to participate in the struggle. Claims to having listened and passed along the news symbolize participation for those who stayed behind. Memories of *Angola Combatente*, crystallized in the dominant trope of clandestine listening, may be where the PIDE's imagination of political mobilization and the MPLA's imagination of participation meet. Still, that trope hides what is most potent about radio, what troubles state power: the specificity of radio, how transduction can act on the body, and audience listening techniques. Those potencies of radio would eventually nettle the newly independent state. Chapter 4 tells that story. To understand it fully and see what the MPLA inherited in 1975, we need to first think about the colonial state and military's counterinsurgency program and how it used and built up radio.

3 ⟿ Electronic Warfare

Radio and Counterinsurgency, 1961–74

> *Is all the enemy's propaganda false? If it were, it wouldn't have any credibility with the people. The colonialist has gotten over the initial phase of stupidity when everything he said was ridiculous. Today they have learned, lie in the abstract with true things in the particular.*
>
> —Pepetela, *Geração da Utopia*[1]

THE CHARACTER "Mundial," a nom de guerre, comments on the successes of colonial military propaganda and the material attractions of surrender. "Ridiculous"-sounding radio (what one Portuguese radio professional described as "too official") matured over the course of the war. Guerrilla broadcasters of *Angola Combatante* and *Voz de Angola Livre* challenged colonial rule, reframing it with a fresh political perspective. The EOA refuted the liberation movement claims with flat, reactive counterpropaganda. In the last years of the colonial regime, military counterinsurgency experts, secret police, and broadcasters finally agreed on a different strategy. They employed Angolan announcers, and broadcast local music and culture to engage audience attention. That is what Pepetela's narrator called "[lying] in the abstract with true things in the particular"—the colonial lie, whether a story of development or lusotropical racial harmony or nationalist movement battlefield desertions, enlivened with authentic Angolan voices, music, and stories.

Guerrilla broadcasting by the MPLA and the FNLA unnerved. It fueled paranoia about foreign broadcasters: hostile forces and voices broadcast from across the territory's borders and locales off the continent that questioned Portugal's continued presence on the African continent. Sonically under siege, nervous, and rattled, the colonial state needed to broadcast its own message. It could not speak with one voice, however. The state's interests fragmented along different lines: the Portuguese National Broadcaster (EN), the military, and the PIDE had different ideas about propaganda. Military interests often prevailed. Military counterinsurgency dovetailed with development and urban planning aimed to combat anticolonial activity and sentiment. The result was twofold: a "pscychosocial action" program that sought to win the "hearts and minds" of Africans and Portuguese in the territory and a bumptious midcentury modern radio edifice. Both used material construction to distance anxieties: development to conquer loyalty, and concrete to fix, literally, that pesky, ephemeral hissing through the air. The immaterial elements of "psychosocial" action, namely propaganda broadcasts, sometimes called "electronic warfare," on the Angolan-hosted *Voz de Angola* (Voice of Angola) proved most effective.[2]

This chapter studies the colonial counterinsurgency project as it touched on radio broadcasting and infrastructure. It exposes conflicts between different colonial institutions regarding whether to prioritize infrastructure over propaganda content or vice versa. Introducing their ideas at the 1968–69 General Commission on Counterinsurgency, then seven years into the war, military brass presented a strategy that required cooperation and coordination between discrete government units and private organizations: the military, the PIDE, and state and radio club broadcasters. Approaching the counterinsurgency project through radio allows me to amplify the dissonances in practices, the "tensions of empire," and the immaterial elements of modernist state projects.

One military official summed up the political situation this way: "Revolt starts where the road ends."[3] Portuguese military generals packaged nationalist politics, development, and the new counterinsurgency plan in a tidy bundle. State radio broadcasting in Angola gained salience as a counterinsurgency strategy, part of the sound and structure of militarization, development, and urbanization. If it sounded "ridiculous" at first—all lies, by the early 1970s it sounded more convincing and attracted attention with true lies.

After 1961, the Portuguese political agenda prioritized war. An emphasis on war bolstered the military's role in society, and it blurred the lines between civilian and military institutions and practices in daily life. Developments in radio embodied that ambivalence. In radio, counterinsurgency strategists sought a technopolitical solution to the problem of nationalist movements and anticolonialism. Given the fractured nature of the colonial state and the sophisticated listening practices of diverse audiences, this technopolitics foundered; implementation of plans to extend radio broadcasting met with limited success and produced unexpected ends. But failure leaves an imprint.[4] The association of state broadcasting with the news cycle (and the apportioning of the day), the attempt to craft effective propaganda, and the Official Angolan Broadcaster's efforts to cultivate different audiences were some of the immaterial ways the various fragments of the late colonial state shaped the soundscape and daily rhythms as much as the urban space of the Angolan territory, even when their plans did not succeed as they had hoped.

RADIO, ELECTRONIC WARFARE, AND PROPAGANDA

The state of war rearranged priorities. It intensified the need for a territory-wide communications system. War also multiplied the list of players interested in broadcasting. What had been slow, rumbling worries, and a clutch of concerns over the need to make broadcasting reflect the new policy of overseas provinces in the late 1950s, turned into a cacophonous din of a problem when the war broke out in 1961. Suddenly radio mattered even more to intelligence collected by the infamous PIDE and counterinsurgency organized by the military. The colonial state´s police apparatus intensified surveillance to record the impact of foreign-based broadcasters and the local use of radio. Surveillance and what they found made military and police nervous (a process addressed at the microlevel in the last chapter).[5] The colonial state needed to broadcast its position, persuade listeners to join its side, and counter nationalist propaganda. This set up a pattern: repeated tensions between military, police, and radio professionals over the question of "electronic warfare" and the infrastructure required to make it possible.

Legislation sketches the shifting institutional stakes of radio broadcasting, while technical reports and studies outline the arguments for change.[6] Early in 1961 (two short weeks after the attacks on Luanda's prisons), the *Official Gazette*, noting the "pressing need" for action in developing telecommunications, increased the "Communications and Transport—Telecommunications" budget to 15,000 contos. Budget increases for another six areas of development combined came to the same amount, heralding a significant commitment to a broader communications footprint.[7] In late May 1961, three months after the anticolonial war had begun in earnest, the state created the Coordinating Commission for the Plan for Angolan Radio Broadcasting (CCPRA) to funnel expertise, not just money, to radio communications.

Answerable to the governor-general and under the broader institutional guidance of the CTT, CITA, and the EN, the CCPRA understood radio as propaganda and the problem as one of range and message. Created as a temporary body, the CCPRA existed until the 25 April 1974 military coup in Portugal. It was first run by three civil servants, two from CITA and one from the Fazenda/Ministry of Finance. A structural revision in 1964 added the CITA director and an electrotechnical engineer from the PRA to the commission.[8] This signaled the importance of technology to the commission as well as the growing significance of radio. The CCPRA aimed to develop a territorial plan for radio broadcasting, organize a public bidding process for the equipment and materials required to execute the plan, and, upon conclusion of its activities, deliver the plan to CITA, an office under the aegis of the PIDE.[9]

The commission and budget emitted a new frequency of power. Instead of just violence, the commission provided a new structure and way of approaching the war. This was a frequency in the sense of a more robust form of broadcast power—more consistent and with a broader institutional base. It emerged at the interface of police, radio professionals, and the military. Yet this new frequency faltered along those very lines of intended cooperation.

No singular written plan exists. Instead, the "plan" was a long-term, shifting set of goals, grounded in infrastructure and targeted at increasing the EOA's broadcasting range. Lacking in trained personnel and a substantial budget, the CCPRA built on an earlier study written by Manuel Bívar (technical director of the EN) and the information about radio clubs collected by the CTT in the mid-1950s.[10] The CCPRA's plan set out to build infrastructure and to remedy the problems of coordination, broadcast range, and training that the earlier study identified and that the outbreak of war aggravated. If records are any indication, much of what the commission did was to listen in to foreign radio audible in the territory and then debate the merits of jamming and counterpropaganda.

The Ministério do Ultramar (Overseas Ministry) issued a flurry of decrees. One removed customs charges applied to imported radio apparatuses for use in the official network.[11] Another extended the radio subscription tax exemption to provincial (i.e., of the colonial territories) governors, secretary-generals, provincial secretaries, and soldiers stationed in the overseas territories, making information available to Portuguese working for the colonial state. A third, from late 1962, decentralized control, shifting fiscal and policy oversight from the National Broadcaster (EN) in Lisbon to provincial governors.[12] Nine months later, official broadcasting in Angola moved from the

CTT to CITA and received its own budget, as foreseen in the CCPRA.[13] This subordinated radio to a department of intelligence and security.

Broadcasters at the EN took the opportunity the war presented to argue for a renewed role of radio broadcasting as a cornerstone of national defense and development. They echoed the need for infrastructure, secondary sector development, and European integration that figured in the two National Development Plans (1953–58 and 1958–64). State propaganda, as an iteration of state power, should produce more than repress, in this vision. Thus the EN's technical director, Bívar, emphasized an "energetic response" to hostile foreign broadcasts in place of jamming and guidelines for national propaganda over the heavy-handed use of censorship's blue pen.

An untitled report from the EN written in 1961 elaborates on Bívar's urging to think of radio broadcasting as an essential part of national defense. "The Portuguese Radiophonic Problem," part 1, argued for a bolder broadcasting system and propaganda plan. In it, Bívar suggested that radio strength is independent of a country's size: tiny Portugal could broadcast with as much force as the BBC, Radio Moscow, RTF, or the USA.[14] Radio, he noted, is more effective than military matériel: "Two or three short wave 250 KW transmitters that work 24-hours a day for national defense can be more effective than an anti-air battery, a plane, or tank, which might not work and costs two to three times as much."[15] While this position aligned with the developing military mind that emphasized counterinsurgency and psychological operations, it arose from longer-term conflicts within the Salazar state over the role of propaganda, radio, and the EN's budget for such work. Like Bívar's earlier report, this one argued that effective, combative propaganda is messy. The official niceties and diplomatic cautions of broadcasting done with the state's imprimatur tend to neutralize propaganda, resulting in "ridiculous"-sounding radio. Rather, the state should draft guidelines to create a clear and consistent message across the different stations.[16] Bívar advocated the formation of a federation and central coordination service run through the state instead of broadcasting done by the state.

Salazar, unlike Hitler, paid little attention to radio.[17] A hot medium, radio mobilized and engaged listeners. Salazar, instead, wanted a quiescent population. His propaganda cultivated passivity; it "focused on praising the head of government's work and the heroic feats of the Portuguese nation, creating a climate of serenity and trust, rather than one of exaltation and mobilization."[18] That said, he did not have a focused propaganda strategy, an organization dedicated to propaganda, or resources earmarked for the question. This frustrated generations of propagandists and broadcasters who bemoaned the effete state of broadcasting: "suffocated by a bureaucratic system, useless due to lack of leadership, and with sparse resources at its disposal," it was overly dependent on the state.[19]

The first part of the report sketched the general problem, and the second part, the "Plano de Radiodifusão de Angola" (PRA; Angolan Plan for Radio Broadcasting), offered a program of action.[20] Given the events in early 1961, the focus turned from the Portuguese opposition to the "terrorist movement," fed by the ever-expanding services of foreign powers broadcasting in Portuguese and local languages (Radio Moscow, Cairo, Brazzaville, Ghana, for example).[21] The PRA's budget, first set at 17,000 contos, was increased to 35,000 contos (equal to the budget of Portugal's EN).[22] Despite the alacrity of the commission in drafting the plan, four months hence, the state had not yet authorized the proposed budget. In parallel, the military and police began to approach radio broadcasting from the angle of counterinsurgency.

THE MILITARY AND COUNTERINSURGENCY

The Portuguese military started thinking about counterinsurgency in the late 1950s. Clinging to the fringes of postwar Europe, Portugal entered NATO and reinvented itself.[23] The Portuguese modernized their military—adopting an improved and more efficient infrastructure—and outfitted troops with new arms as NATO required. For Miguel Bandeira Jerónimo and António Costa Pinto, this was part of a larger set of changes they call "repressive developmentalism."[24] Changes in military structure and operations and more broadly cast institutional and policy changes aimed to join development, security, and governmentality to modernize Portugal and its colonies.

In this post–World War II, Cold War context, integration into NATO emboldened the anticommunist elements of Salazar's military. Rumors of international communist plots to foment revolution in Portugal's African colonies inspired military brass to draft plans for counterinsurgency. They anticipated rebellion. They studied the extant experiences of Western states fighting guerrilla forces—the British in Kenya and Malaya, the French in Indochina and Algeria, and the United States in Vietnam.[25] The British and French experiences proved particularly apropos (*CiA*, chap. 3).[26]

External communist agitation served as a leitmotif, consistent with the geopolitical trends of Portugal's allies. For example, in the late 1950s, Portuguese military captains studying in England translated a British training manual on the communist roots of insurgency. John P. Cann notes that the military "did not acknowledge that nationalist movements might be motivated by a simpler and more straightforward desire for independence, and the lack of this aspect reinforced the Portuguese theory of a communist conspiracy" (*CiA*, chap. 3).[27] Cold War ideologies distracted from clear-eyed assessments of the effects of Portuguese politics and policy in Angola, Mozambique, Cape Verde, and Guinea-Bissau.

By the late 1950s and into the 1960s, the Portuguese military developed psychological tactics and planned civilian development projects. Cann calls this the "Portuguese way of war" (*CiA*, subtitle), but Christopher Day and William Reno describe it in less singular terms as the population-centric strategies of classic counterinsurgency.[28] Counterinsurgency aimed to be low in cost and slow in tempo (thus viable in the long-term). Soldiers would protect Africans in the "overseas provinces" from the nationalist insurgents, redress grievances, and improve the standard of living without straining support in the metropole (*CiA*, preface). The colonies would bear the financial and material burden: more African troops and budgets drawn from local revenue. At the same time, a good deal of the defense budget would go to shore up social programs (education, health, agriculture) so long neglected (*CiA*, chap. 1 and 5).[29]

Work on counterinsurgency began before the war broke out in 1961, but fighting in Angola put it to the test (*CiA*, chap. 1). Brazen military bombing in February and March in Angola's east, the mobilization of civilian forces in Luanda during February, and the use of terror in response to the 15 March attacks in northern Angola were a stuttering if resolutely violent retort to the uncoordinated acts of African resistance. The military could airlift only a small number of troops initially. The military delayed formal occupation until mid-May and full deployment waited until June, when thousands of troops arrived by sea (*CiA*, 16). White settlers strained against the military imposition and demanded more autonomy.[30] Fighting wars in three territories (Angola, Guinea-Bissau, and Mozambique), the military tweaked tactics and adapted to events on the ground, experimenting in action, especially in Angola. They published official doctrine in 1963: "O Exército na Guerra Subversiva" (The Army in the Subversive War) (*CiA*, chap. 3).[31]

Psychological operations (propaganda and counterpropaganda) formed a central plank of the "subversive war" (*CiA*, chap. 1).[32] Salazar's economic austerity restrained the military budget and forced the military to turn to other strategies, like *acção psicologica* (psychological action).[33] Guerrillas used the strategy as well, if not with the same name. Psychological action aimed to win the support of civilians, demoralize the enemy, and maintain soldier morale.[34] For the army, psychological action, psychosocial action (development), and the action of presence (boots on the ground) worked together. The military began implementing this plan in 1965, just as the broadcasts of the MPLA's *Angola Combatente* began to sizzle across the airwaves.[35]

Counterinsurgency activities blurred military and civilian worlds. Military structure shifted. By the 1968–69 Symposium on Counterinsurgency, the remit of the General Council of counterinsurgency bled into governing: "Given that the activities of counterinsurgency are not easily distinguished from those of Public Administration, the GC should act as 'the General Government and the Military Command Council.'"[36] Counterinsurgency became the day-to-day of colonial administration. In practical terms, this demanded coordination between military, police, and civil administration. Police units predominated in cities while army units took charge in rural areas. If in the metropole, one historian describes the PIDE's labors as an example of dictatorial repression (they went after Salazar's opposition, not after everyone), in the colonies, their efforts might best be described as totalitarian repression (everyone was a potential insurgent).[37] Counterinsurgency struck broadly. Or, in the words of the introduction of the counterinsurgency symposium's final report, "Activities of counter-subversion are not easily distinguished from activities of public administration."[38]

Intelligence and military priorities infiltrated all aspects of life. Secret police surveilled the commercial and cultural activities of average residents. The PIDE opened and read the foreign correspondence of musicians; stymied commercial outfits' ambitions to sell electronics equipment in the interior (they might set a "precendent" for groups with less illustrious credentials and less noble aims); and obstructed the folkloric group Kissueia's (where internationally famed musician Bonga got his start) entreaty for stage costumes, squeezing and twisting it through countless layers of bureaucracy.[39] Files full of requests to follow up on proposals, earlier memos, and letters testify to how slowly the system worked.[40]

The secret police produced the very enemies they set out to find as they tailored the world to their nationalist-fascist vision. Ruy Blanes shows that police actions aimed at containing Tocoism catalyzed a regional phenomenon into a territorial one.[41] The counterinsurgency plan had a similar impact. On the one hand, the military directed resettlement in *aldeamentos* (strategic resettlements) aimed at self-defense and military organization in areas where the nationalists organized (especially in Angola's east from late 1966 on) and, on the other, the civil administration fostered rural development in what was called *reordenamento rural* (rural resettlement) in noncombat areas.[42] In both cases, the military concentrated populations along transportation routes to facilitate infrastructure construction, service delivery, and patrol. In the strategic resettlements Gerald Bender analyzed, he found that the military refused to arm the village defense militias and often dispossessed Africans of land for Europeans, reproducing the racism of colonial rule and fueling anticolonialism.[43]

The military and police strategies for winning the war comprised conflicting tactics. This distilled into a compromise: a twin-pronged approach of control and development, sometimes articulated as violence and reform.[44] Bender describes the effects of resettlement: "While resettlement failed to stop or even attenuate the insurgency, the radical changes it provoked in traditional society were so profound that rural Angola will never be the same."[45] Daniel Branch reached a similar conclusion in his analysis of British counterinsurgency in Kenya; the violence in these so-called nonviolent programs of relocation and their long-term effects undermined trust just as the Africanization of troops created political fallout and often violence.[46] Like the rest of the Portuguese counterinsurgency program, British, French, and US programs inspired these efforts. That their effects were similar is not so surprising. That they continue to be held up as models for contemporary counterinsurgency, however, is troubling.[47]

Military counterinsurgency prioritized resettlement when the nationalists changed their tactics and opened up the eastern front in late 1966.[48] In eastern Angola, the MPLA and UNITA, which Jonas Savimbi formed in 1966 after he split with the FNLA, established political camps and military bases that allowed them to control territory, harass colonial troops, and access local populations. The Portuguese military believed that resettlement areas would keep the local populations away from the dangerous nationalists, even as they produced the very effects they sought to eliminate. Meanwhile, the number and quality of Portuguese soldiers declined. The strain of the war made itself felt in the metropole, while "hostile" foreign broadcasting rippled across the Angolan territory.

BUILDING EMISSORA OFICIAL DE ANGOLA: URBAN DEVELOPMENT AND RADIO POWER

How did this matter to radio? The same problems of coordination, conflict, and state fragmentation beset plans for broadcasting and for surveilling foreign broadcasters. Hostile foreign broadcasting, and the nagging vituperations of nationalist guerrilla broadcasters, crowded airspace and crackled on transistors. SCCIA, CITA, the PIDE, and now the military all expressed opinions about broadcasting. But as the debate about jamming versus content continued, "the Plan" (PRA) took on concrete form, quite literally, in the construction of a new radio building. Counterinsurgency and urban development intersected in a technopolitical plan. The state sought technical and technological solutions for the political and military problem of nationalism: radiocentric, modern urban growth and maximum kilowatts for greatest range housed in a new, modernist building in the capital with powerful new transmitters at the Mulenvos Transmission Center.[49]

Repeating Cold War ideologies of the period and of the militaries with which they had trained, the PIDE, military counterinsurgency experts, and radio broadcasters believed that the infiltration of foreign, communist propaganda instigated nationalism. They imagined technical solutions: jam or counterbroadcast, increasingly the latter. The Overseas Ministry invested more in infrastructure than in personnel, in bricks and mortar, transmitters, and cables than in training, suggesting a commitment to technical solutions. In this respect, radio broadcasting differed from counterinsurgency, which was less technical and more soldier-centered, less about guns than human relations (see *CiA*).[50]

Construction of the new EOA in Luanda took four years: 1963–67. The building embodied "the Plan," its contradictions, and the state's investment in technical solutions to political problems. It was the only radio station in Portugal built from the ground up solely for the purpose of radio broadcasting. It boasted the newest technologies and the most advanced sound equipment in territorial Portugal. As a piece of state architecture, it was a midcentury modernist edifice built to project colonial state authority.

Entangled in the authoritarian modern state-making practices James C. Scott describes as "seeing like a state," the radio station was a visible, physical manifestation of state power. Radio professionals remember that even after the building's construction, they continued to refer to it as "the Plan," such was its hold on the imagination of radio employees. It represented the state's power in brutalist, hard-edged, concrete terms and also its capacity to build. In terms of radio propaganda, it left much to be desired. It sounded far too much like a state.

The institutions associated with the colonies and with the war (the military, the secret police) took modernizing strategies of legibility and institutional coordination most seriously.[51] The war and a Europe reconfigured after World War II shaped the Portuguese state's desire to insert Portugal into a new international regime that centered on information, developmental colonialism, and norms regarding imperial and colonial rule.[52] The new international colonial lexicon took a technocratic and scientific turn, emphasizing terms such as "planning," "good government," and "efficiency."[53] Internal, domestic motives existed too. The state might avoid costly purchases of military matériel if it modernized administration and built infrastructure.[54] Despite the failures of the National Development Plans, they received more of the national budget than did the military, even during the war.[55] Development doubled as counterinsurgency.

Efforts at economic reform in continental Portugal followed the Marshall Plan and the country's bid to integrate into Europe. Some advisers advocated that Portugal detach from the colonies entirely.[56] As such, the First Plan (1953) targeted foreign investment in the economy at only 6 percent. However, by the late 1950s, the Second Plan for National Development (1958–64) promoted greater integration between Portugal and the colonies. With conflict brewing on the horizon in the colonies, and European integrationist technocrats occupying more seats in Salazar's cabinet, foreign investment goals rose to 25 percent. The secondary sector took priority, marginalizing what had always been an agriculturally driven economy. With the Third Plan (1968–73), foreign investment benchmarks reached 34 percent.[57] The National Development Plans generated a new politics of public works in the 1950s and 1960s. The state built transport networks (roads, railroads, airports) and energy-related infrastructure (like dams) on the African continent in this period.[58]

This new infrastructure in the African overseas territories, especially Angola, aimed to lure and redirect white migrants there. A political ploy, too, this would show the world that Portugal was in Angola to stay. White settlement meant to develop the economy in Angola and implant the Portuguese in the territory: "Angola é nossa!" (Angola is ours!), they proclaimed.[59] The white population in Angola increased by 87 percent in the fourteen years of the war.[60] Urban development and infrastructure—roads, telecommunications, housing—and industry all grew, but rarely kept up with immigration, as the surge in illegal construction indicates.[61] Counterinsurgency by settlement dispossessed Africans of their lands.[62] It made them more insecure and fueled rural-to-urban migration.[63] Bandeira and Pinto put it succinctly: "If the civilizing mission was a mechanism for engineering inequality, welfare colonialism was in essence a form of managing it."[64] Managing inequality was not the same thing as correcting it.

In the midst of the war and growing urban populations, city planning plotted to organize space for optimal control.[65] António Tómas argues that radiocentric urban development, building on Le Corbusier´s Athens Charter, attempted to impose a modernizing control on burgeoning Angolan cities from the 1950s. Modernism emulated colonialism, destroying the old (African) to build the new (European) city, spreading outward from a "central nucleus," structuring and administering social, political, and economic asymmetries based on racial inequalities along the way.[66] The colonial state did not succeed in fully implementing any of these plans, but the plans still shaped urban space and development.[67] Even if indirectly, urban development was counterinsurgency. Developing industry, employing Africans, and opening schools and health centers created a material bulwark aimed to counter the argument of African nationalists.

It did not always work. In Luanda, the Gabinete de Urbanização de Câmara Municipal de Luanda (Urbanization Office of the Luanda City Council), opened in 1961 under the direction of architect and urbanist Fernão Lopes Simões de Carvalho. The office generated plans for urban industrial development, housing, and recreation in the *Plano Director* 1961–62 (Guiding Plan).[68] This plan was built on a wide set of studies (on water and electricity provision, sewage, population needs, racial integration) Carvalho undertook with sociologists and architects in his office. But the City Council, in charge of urban growth and development, was notoriously corrupt, and urban development became hamstrung between state plans and private interests.[69] The plan never made it to the City Council's assembly for approval, stymied because the City Council president's brother-in-law was an architect whose plans the president promised to promote.[70] Development thus benefited Portuguese immigrants both in terms of the contracts tendered and in terms of the kinds of plans and projects approved. Inequalities between locally born Portuguese and those from the metropole, not to mention institutional imperatives, exposed the tensions within and across empire and the fragmented nature of colonial state interests.

Nonetheless, Simões de Carvalho undertook a number of projects.[71] He was among three in a generation of architects (including Vasco Vieira da Costa and Nadir Afonso), trained in Portugal and known as tropical modernists, who studied with Le Corbusier.[72] His work is deeply marked by Corbusian principles and a brutalist style—raw concrete, visible brickwork, rugged construction, large volumes. He parted ways with Le Corbusier on the question of working from a tabula rasa and, unlike Le Corbusier, was committed to understanding the needs of those meant to reside in his buildings. He was then (and is now) a critic of Estado Novo racism and the corruption of Portuguese bureaucracy. He sought to use building design to advance urban development.[73] Whatever his differences with colonial policy, Carvalho is still a good example of the ways postwar knowledge production, in the language of science and rationality, suffused new state practices.[74]

Carvalho designed a television station for his final thesis project in architecture. When the colonial state opened a public tender for the design of a radio station, he and architect José Pinto da Cunha put in a bid. Their proposal won, and Manuel Bívar and Humberto Mergulhão, in charge of "the Plan," sent Carvalho on a trip to visit state-of-the-art radio stations in

Europe—Paris, Amsterdam, Strasbourg, Hamburg, and Baden—demonstrating again how international models served to shape Portuguese modernization practices during the Cold War. On this trip he learned that the two main architectural concerns for radio stations were (1) security; and (2) phonic isolation. The newest, most advanced stations built subterranean studios to create the best sonic conditions.[75]

In Luanda, the EOA/RNA building consists of a large, rectangular main floor organized around a longitudinal corridor, with internal, open-air gardens, marble halls, and wood-paneled sound studios. A subfloor contains warehouses, technical workshops, and an auditorium. The facade is concrete with alternating horizontal and vertical lines creating *brise-soleil* for cooling.[76] Vegetation planted on the roof would provide both cooling and sound protection from the nearby airport. Transversal air vents facilitated natural cooling through most of the building, while air-conditioning was reserved for closed studios and workrooms.

This, then, was a midcentury modernist counterinsurgency project that would make a claim on urban space and territorial sound. The planning offices of the city drew up urban housing for the masses, reorganizing urban space in a modernist and racist design in an attempt to control urban populations and to justify continued Portuguese presence under the guise of development.[77] Centered, literally, in the expanding city, amid the new modernist Sagrada Família (Sacred Family) Catholic Church, a military hospital, and the airport, the new EOA figured as part of an urban communications and command hub with territory-wide ramifications.

What stands today as Rádio Nacional de Angola (Angolan National Radio), the erstwhile EOA, is an incomplete version of Carvalho's plan. The second phase, a five-story administrative and social services building (offices, a school to train journalists and technicians, and a daycare for workers' children), as well as extensive landscaping, never materialized. The government of independent Angola initially contracted Carvalho to finish the work, but the war eventually distracted priorities and budgets.

The building, as it stands, is likewise incomplete. Nonetheless, it is an impressive and imposing edifice. Carvalho designed studios overlooking gardened patios because he did not like the idea of announcers being in "buried zones," underground. Triple-thick glass windows protect against exterior sound intrusion. Carvalho used transparency to solve the security question. That is, instead of closing off the radio station, he opened it up visually. A large glass front, approached from a wide bridge, opens into a light-filled atrium with a shallow pool, next to the main operations office, also housed in glass. He believed that when you could see out, you could see who was approaching.[78]

FIGURE 3.1 Atrium and central operations at EOA after the building's completion in 1967. The pool is just beyond the columns on the right, and the glass-enclosed operations room is on the left. Photo from the collection of Fernão Lopes Simões de Carvalho and used with his permission.

SOUNDING LIKE A STATE

The EOA broadcast news of battle victories, military press releases, programs for soldiers written by the military, promotion of development projects, and propaganda aimed to convince the guerrillas to abandon the nationalist movements.[79] One counterinsurgency report even proposed television as a distraction from radio listening, so attractive did the military imagine the foreign broadcasts to be to African ears.[80] This was "psychological action" at work. Radio clubs rebroadcast the material. Radio journalists from the period recall that only these official state broadcasts featured war as a subject. Otherwise, at Rádio Club de Huíla, and other radio clubs, state censorship and self-censorship ruled out discussion of the military conflict or the liberation struggles.[81]

For some colonial and military bureaucrats, propaganda that lied was a liability. As a result, the EOA was too transparent: it sounded like a state. Bívar pointed repeatedly to the compromised sound of state broadcasting and the successes of radio clubs. Bureaucrats of the CTT and the colonial administration disagreed. They viewed the local, amateur radio clubs and their attractive sounds with skepticism. A 1953 government dispatch to hire staff for the newly formed broadcaster argued that new employees should be trained at the radio station to protect them against the "vices and prejudices" developed at the radio clubs, echoing the differences between metropolitan and territorial interests of the white population.[82] These voices predominated.

They did not, however, prevail. As foregoing sections argue, the war shook up the thinking in some state offices. Elements of the PIDE and the military argued for a nimble counterinsurgency program with a voice that could speak to the "spirit" of Angolans.[83] Another set of critics, mostly radio club adepts, agreed. These critics weighed in on the question of radio and the EOA in the alternative daily paper *ABC* between August and October 1967. By that point, the contents of radio broadcasting clearly fell short of the new building's physical presence, and radio club members and broadcasters debated its merits and limitations.[84]

One writer, Maurício Ferreira, echoed Bívar's concerns about the problematic name and structure of the broadcaster.[85] Now that "the Plan" had given the EOA the technical capacity to broadcast not just to the whole territory but well beyond its borders, its name and sound undermined its legitimacy. As an official state organ, Emissora Oficial (Official Broadcaster), instead of a state corporation like the BBC or the mixed trust Portuguese Radio and Television, was too constrained by bureaucratic process and the repetition of administrative news.[86] It sounded official and hollow; all the dull buzz of technical power without the warm, human hum of programming. People, states, and liberation movements across the continent could hear this failure.

As if that were not enough, pseudonymous writer V.S. lamented the prejudicial financial situation of the EOA and Angola's radio clubs relative to the metropolitan EN: "Fifty percent (which translates to many thousands of contos) of the radio broadcasting taxes collected among us go to the coffers of the EN . . . when they should be distributed to the EOA and other provincial broadcasters as compensation for the many dozens of hours of broadcasting done annually for the government."[87]

The author went on to criticize the fact that when EOA professionals went to work at the EN, they continued to receive their salaries from the EOA budget, but when EN professionals came to the EOA, the EOA paid them directly, gouging colonial budgets to the advantage of metropolitan ones. The radio in the colonial territory paid a heavy price in terms of tax revenue and autonomy.

Private correspondence echoed with similar concerns articulated in different terms. A 1967 letter from Rádio Comercial, found in the GNP archives, alerts the state to the radio's ambition to expand its range. The letter is framed first by its author, Eurico Mota Viega, one of Rádio Comercial's investors. Viega argues that a commercial radio station is best situated to broadcast counterpropaganda. It could relieve the state of having to subsidize smaller radio clubs, redirecting state money to development and defense; it could defend against "subversive agents" entering radio given their excellent technical and administrative capacity; and it could provide wide coverage in medium wave to be an effective messenger "for higher national interests" by installing a 1 kW medium-wave transmitter in Luanda and a 10 kW one in Sá de Bandeira.[88] This would provide the government with an unofficial voice for counterpropaganda (a private radio cover for a state motive) to be broadcast inside the territory and to neighboring countries.[89] Even though Rádio Comercial boasted the technical expertise of Sebastião Coelho, formerly jailed, they hoped his and their whiteness and their explicit support for "the Nation" would convince the state.

SEEING LIKE A COUNTERINSURGENCY STATE

Sounding like a state developed out of seeing like a counterinsurgency state. But visual regimes did not translate so easily into aural ones. In 1967 the colonial government formed the General Council for Counterinsurgency and convened a symposium from November 1968 to March 1969 that included top members of the army, police, and information services.[90] They produced twenty-five secret reports for discussion.[91] The main areas concerned the general plan and organization of countersubversion, the regrouping and control of populations, civil defense, social promotion (namely, rural resettlement schemes), and the psychological action and public information projects.[92]

The General Council maintained its own "bodies for decisions, actions, information and orientation" within Angolan administrative structures.[93] Every district, municipality, and locality organized counterinsurgency councils to foster intimacy with "the desires of the African population[,] which we must keep and withdraw from the enemy," in the words of the report.[94] Counterinsurgency comprised military, political, and psychological action targeted at controlling the populations.

Despite discussing a voice that resonated with the "Angolan spirit" and "African desires," a spatializing, visually driven legibility predominated. The documents from the symposium on counterinsurgency, as detailed in the

"Final Conclusions" and "Findings Report" of the symposium from the military archives, summarize the military and political plans implemented to that point and those projected for the future. The growing presence of nationalist guerrillas inside the territory and along the extensive and permeable eastern border (from 1966 on) required an urgent response. Plans for resettlement, whether for military ends or to facilitate development, always entailed information gathering. In the aldeamentos, a census enumerated the population. The colonial state issued a family identity document that listed all the inhabitants of a home. One copy stayed in the home and another at the local antisubversion committee. Anyone unaccounted for during a check (without the requisite permit for absence) was declared an enemy collaborator and the police issued a warrant for the missing person's arrest.[95] A section on the difficulty of implementing resettlement with pastoral populations (and the resultant higher requirements for radios, cars, and personnel) in Huíla district underscores how much fixity and legibility mattered to the military and police.[96]

Another section covered the "Control of the Population by Means of Identity Cards." Lieutenant Colonel Fernando Lisboa Botelho, head of the second General Council on Counterinsurgency, argued that a new law would institute a "modern" system of identity based on residence, replacing the prior system based on names: "By decentralizing the archives, distributing them over the various police districts and giving each policeman a specific area, it will be possible to gather the data indispensable for an efficacious checking of persons who are suspected of subversive activities."[97] Counterinsurgency experts cataloged a number of added advantages: easier census-taking, greater knowledge of Luanda's population over age twelve to help determine eligibility for military service, tax collection, election participation, potential food rationing, and supporting the information system.[98]

These activities fit James Scott's descriptions of states mandating modernization and using bureaucratic techniques that promote legibility to administer and control growing populations.[99] Such states sought to foster economic and social development through rational planning, implementing standardization, and producing comprehensive knowledge about populations and national territory. They promoted well-being in the name of modernization but kept oversight and control paramount. If the sound of guerrilla broadcasters induced nervousness, visual accounting might produce calm.

From the point of view of the PIDE, liberation movement propaganda, specifically radio, posed the greatest threat. Radio could mobilize new recruits and destabilize the African populations. The PIDE's reports to the counterinsurgency symposium proclaimed the success of guerrilla radio. PIDE officer Manuel J. Correia reported:

> The radio programs of the enemy are extremely well prepared. They are broadcast by foreign stations, such as those in Brazzaville, Kinshasa, Tanzania, Ghana, Prague and Moscow. They influence the population to such an extent that it takes to passive or even active revolt. Following the radio propaganda, they become dangerous agitators. By forming cells, which in their turn grow and then split up again, bases come into being which are essential to the progress of the terrorist troops crossing the border and forcing us to irregular warfare without fixed fronts.[100]

The metaphor of organic, cellular growth (and indeed, the guerrillas used the concept of cells, but they were small and far-flung), proliferating and producing illegible ("irregular" and "without fixed fronts") forms of fighting, aptly expresses the anxiety of a police force and military that could no longer discern or contain the enemy. Radio waves—invisible, amorphous, out-of-control frequencies—from widely dispersed broadcasters likewise refused the kind of fixities that police, military, and colonial states needed to stop the insurgents. In this sense, radio was not an obvious choice for counterinsurgency. Much of what the military and the police focused on, well into the late 1960s, centered on jamming hostile broadcasts. That was a tangible, defensive action.

SOUNDING "AFRICAN": *VOZ DE ANGOLA*

One propaganda station proved popular, if not entirely successful. *Voz de Angola*, broadcast from a structure alongside the EOA, was the semiofficial sound of counterinsurgency. It had a questionable juridical status as late as 1972, which kept it not quite a department of EOA. This hazy status—peripheral to the state broadcaster but not legally existing—made it, by default, precisely the kind of station Bívar proposed. The commission discovered the legal discrepancy of the station's status by chance. In response to a questionnaire sent by an academic who was writing about radio in Portuguese Africa, CORANGOLA decided it needed to have a set of pat answers about the station.[101] In formulating a response, they realized that the station did not exist legally. The result of a secret dispatch from the ministers of national defense, the interior, the overseas territories, and the secretary of state for the President's Council, CORANGOLA had the responsibility of producing "psychological action" programs. None of the other radio stations, private or public, had the staff capable of assuming this work. According to the note, "In the eyes of a great majority of our public, it is seen as a private station. And it will serve us if people keep thinking so."[102]

Voz de Angola broadcast true lies, resounding with authentic, local cultural sounds to support a mendacious politics. It had an audience among Africans, some of whom wrote to celebrate and encourage it.[103] It offered a variety of programming, largely with African announcers.[104] Sometimes it made people dance; sometimes it spoke in the voice of captured soldiers and sometimes in local languages. Very rarely was it mistaken for anything but what it was: propaganda. Unlike other programming, consistently described as too staid and official-sounding, *Voz de Angola* sounded more local and dynamic. One PIDE officer described the enthusiasm of listeners this way: "These days no one resists the temptation to tune in to 'Voz de Angola.'" Those at first resistant had been convinced and found "the idea so genial that some doubted that it had sprung from a Portuguese mind."[105]

Some people who wrote to congratulate *Voz de Angola* responded specifically to the broadcasts in Angolan languages, suggesting some corrections or offering their own services in the process.[106] One asked for a music request show on the Tchokwe language program.[107] A handful of the letters came from soldiers and schoolteachers, classic "middle figures" with a stake in the colonial system and its ideology. Other writers celebrated *Voz de Angola* and Portuguese colonization in general.[108] This post-1961 moment of investment and supposed equalization—the removal of the *indigenato* laws, the expansion of education, and the extension of Portuguese citizenship to all—would have been particularly appealing to this social group. Some documents record programs and letters from local authorities who cooperated with the colonial regime and sang its praises on *Voz de Angola*.[109] PIDE officers communicated a general sense of audience favor in reports and responses to an inquiry about the possibility of covering football matches in the territory.[110] Responses from Luso (Luena) in the east, São Salvador (Mbanza Kongo) in the far north, and Salazar (N'Dalatando) in the center reported vibrant listening publics, sometimes both African and European, and enthusiasm for the bilingual broadcasts (in Portuguese and a local language) of football matches, though all requested that metropolitan games be broadcast too.[111]

The station inadvertently proclaimed the success of the radio war between the MPLA and the FNLA and the Portuguese colonial state. *Voz de Angola* was the closest thing to effective counterpropaganda. It addressed African listeners with broadcasters who spoke Kimbundu, Umbundu, and Kikongo. This station played local tunes; indeed, it spurred musical production and helped spread semba music and dance from Luanda throughout the rest of the territory.[112] In 1968 one PIDE officer, Correia de Lima, celebrated the program's success in combating the dreaded "Congomania": the heightened consumption of Congolese music that signaled interest in Congolese independence. He observed that when "walking around the musseques, not so long ago, one saw blacks [he used the perjorative 'pretos'] that carried transistors tuned in to broadcasters from the Congo."[113] Lima also applauded the radio show "listeners' requests" for its popularity, another strike against "Congomania" and, he imagined, the anti-Portuguese programming from Brazzaville. Another officer emphasized the appeal of the station's music to the majority of Africans with their "innate taste for melodious and agitated music."[114] News on *VdA* highlighted development, Portuguese military gains in the war, and the abandonment of guerrilla forces by individual soldiers suffering from hunger, exhaustion, and low morale. The military found it a great success and charted its popularity among both civilian and military listeners.[115]

One show, *Ponto para Meditação* (Something to Think About), proved contentious to both the PIDE and listeners for raising political issues too directly. The officer Correia de Lima suggested this should be done with greater subtlety through the addition of nonpolitical topics, particularly for listeners within the territory.[116] Another officer, Victoriano, who often reported information that criticized the Portuguese, observed reactions to the content and tone of the program among African listeners. His analysis was unsparing. Among the African majority, he said, people doubted the program's credibility. When it explained terrorist actions; claimed there was no racism in Angola; described Angola as a meritocracy; discussed the privations in the Congo; and accused the "terrorists" of killing innocent civilians and children, listeners' experiences made them skeptical. He explained: "In truth, such affirmations are only convincing to an African who has not been a direct victim of any less than human act on the part of a white person. What will the widows, orphans, the very many Africans who saw their homes destroyed by the fury of reprisals that followed the events of 1961, and even those who were robbed of their belongings by unscrupulous individuals even before 1961, say?"[117] Such lies only made "nationalist doctrines" more attractive.

CORANGOLA and the PIDE thought music could deliver distraction and be the sugar for the propaganda pill. Listeners' letters show that they could disentangle tunes from ideology. The MPLA sent menacing letters to the station: "One day your mess of a broadcaster will disappear. Your propaganda? No one believes it, given that even blind people know you are only friends with black people to exploit them and their land. . . . Read this on your program 'Listeners Have a Word.' If you don't, it just shows your cowardice and insecurity."[118] On *Angola Combatente*, broadcasters called out the names of *Voz de Angola* announcers whom they considered to be *bufos*, or informers for the PIDE.[119] True African voices on *Voz de Angola* were liars and traitors in this reading.

Other listeners heard more lies than truth. They wrote in to complain of *Voz de Angola*'s and the state's duplicity. "Why do you promote *Voz de Angola* and then arrest people found with radios?" one João Kahombo from the central highlands city of Nova Lisboa queried in 1969.[120] Another listener wrote to the program *Sanzala de Paz* (Village of Peace) to say that the program discussed what was going on in other countries like Tanzania or Uganda and never mentioned the way whites mistreated "the owners of the land, Angolans," a regular occurrence in some of Luanda's neighborhoods. He noted that everyone knew *Voz de Angola* was part of "Psico" (psychosocial action). "Your time is coming," he warned and signed off: a future engineer from the University of Angola, 1972.[121] Still another letter, also from 1972 and forwarded from the PIDE office in northern São Salvador, written by one António Costa de Campos in Luanda, pointed to the contradictions: on the program *Caminho de Paz* (Pathway of Peace) the announcers preached peace, but "we don´t see anything new, except the same massacres by the colonial regime."[122] Finally, a letter from Francisco José in Luso in late June 1971, addressed directly to Reis Esteves, a Portuguese employee of *Voz de Angola*, asked that his letter be read on-air. *Voz de Angola* tells lies, he said. It encouraged people to come back from the bush to their villages, whereupon the PIDE arrested them. Despite his complaint, he requested that they dedicate a song to him on the following show.[123]

Sometimes *Voz de Angola* responded to listener letters. A letter from 1970 denouncing acts of racism in the southern town of Pereira d'Eça received the following response on the program *Listeners Have a Word*: "You can be sure that, here in the north of Angola, and more so in the metrópole as well, that never happens."[124] You can be sure that many Angolans knew better.

EOA AUTONOMY

In September 1970, the EOA became an independent service that absorbed the former staff, equipment, and spaces previously housed under CITA's shingle.[125] The EOA would expand its staff and its physical space. Building on the findings of the first (and only) colloquium on radio broadcasting in Angola, the Overseas Ministry decided that the Official Broadcaster (EOA) required autonomy in order to effectively fight the propaganda battle.

Meanwhile, all across the decade of the 1960s and into the early 1970s, radio clubs grew. While the state, through the military and the PIDE, used radio as part of psychological action to convince Africans of Portuguese beneficence and commitment to economic development, the PIDE also preoccupied itself with enemy propaganda and enemies in general. The anxieties were palpable as the state listened in on *Angola Combatente* and *Voz de Angola Livre*, and ranged more broadly. Fascism in the metropole and the "overseas provinces" had generated a Portuguese opposition, and not even Portuguese settlers could be trusted. Military and PIDE documents show concern for what the white population was thinking (as evidenced in the last chapter).

Counterinsurgency listened widely and struck more narrowly. Following Pepetela's narrator, *Voz de Angola* lied in the abstract about the wonders of Portuguese rule while telling truth in the particular, articulating an authentic Angolan sound. Counterinsurgency sang in African languages, jammed, built roads and buildings, and sent out soldiers to be poles of attraction for the Portuguese project and polity. Some of those soldiers, if letters to the local press are any indication, wanted very much to stay in Angola. The back-and-forth between metropole and colony, the energies of attraction and repulsion, the frequencies of power, and its infrequencies, too, enlivened unintended networks and desires.

Changes in the counterinsurgency structure followed shortly upon the military coup in Portugal in April 1974. In November, the military proposed a shift. Rather than abolish counterinsurgency while the military continued on the ground in Angola during the transition, the military junta renamed it "civil-military coordination." "Counterinsurgency" suddenly had an "inadequate ring" to it.[126] Finally, the Portuguese state worried about the sound of something. Civil-military coordination mainly concerned itself with "security and economic, social, political, and cultural promotion of the populations, especially in the rural areas."[127] Counterinsurgency as state administration continued apace. Radio broadcasting would shift when the MPLA took over, or would it?

4 Nationalizing Radio

Socialism and Sound at Rádio Nacional de Angola, 1974–92

THE RADIO program *Kudibangela* aired in August 1974 during the period of transition between the 25 April 1974 military coup in Portugal and Agostinho Neto's declaration of Angola's independence on 11 November 1975. *Kudibangela* opened with the call sign "Kudibangela, weya, weya!" and a few bars of a Manu Dibango tune. A Cameroonian saxophonist of world- and continent-wide renown, Dibango's Pan-African funk and jazz sparkled with the energy of independence. "Weya," from the album *Makossa Man* (1974), circulated in the wake of "Soul Makossa" and its highly sampled refrain "mama-se mama-sa mamako-sa." Dibango's "Weya" received less acclaim, but in Angola it not only resounded with the celebratory sentiment attending independence, but it created a linguistic echo. In Umbundu, one of the most widely spoken Angolan languages, *weya* means "arrive." Produced in 1974, Dibango's tune must have sounded like it was welcoming the arrival of Angolan independence.

Kudibangela "sampled" "Weya" long before Michael Jackson, the Jungle Brothers, 7L and Esoteric, Afrika Bambaataa and Soulsonic Force, or DJDay, though its future-oriented, African sound appealed to the show's producers too. In 1975, *Kudibangela*'s call sign buzzed and bumped with the cosmopolitanism of Luanda's *musseques*—resonant with African modernity and volubly Angolan.[1] It struck a national developmental chord: here on the cusp of independence, the show's Kimbundu name, meaning "building;

construction," anticipated the future instead of dwelling on the wrongs of the colonial past.[2] It touted the MPLA party line, calling out the FNLA and UNITA as imperialist puppets, and urged unity, marrying Umbundu and Kimbundu in its call sign. Yet the show ran only until the end of 1975, just weeks after the declaration of independence by the MPLA.

Kudibangela sounded discord in the MPLA. The radio program's dissonance was part of the world of vibrant political debate and critique of the movement-party between 1974 and 1976.[3] The dissidents boasted various ideological stripes. Many shared a position within the MPLA that supported reforms that Bernardo Batista "Nito" Alves and José Van-Dunem advocated. This culminated in the events known as the 27 de Maio and the purge in the MPLA. President Neto and his supporters orchestrated and conducted a party cleansing with long-lasting effect. The fragmentation that characterized the MPLA in this period crackled across the airwaves. Political differences convulsed the radio station. The divisions included not only the supporters of Nito Alves, implicated in a coup on 27 May 1977, but former EOA employees loyal to the MPLA. It sounded out the future of the station.

Although the MPLA prohibited *Kudibangela* from RNA's future, its call sign and sound played on something consistent: innovation. Angolan musicians and Angolan radio broadcasters took up the charge of creating something new, something Angolan (while struggling over what that meant), sounds befitting a modern, socialist nation. *Kudibangela*, even if it struck a discordant political note, also hit on a shared desire to create a new sound and structure at RNA. New sounds and structures are the subject of this chapter. What were they? And just how new were they? How did they create a state radio different from that of the late colonial period?

A December 1975 decree turned the colonial radio into a national broadcaster, but it was the events around 27 May 1977 that consolidated the radio as an organ of the MPLA party/state. It might seem inevitable, given the success of a guerrilla broadcaster like *Angola Combatente*, and the way it made the colonial state bristle, build, and bellow, that the MPLA would turn the EOA to its own purposes without missing a beat. It did not. The Ministry of Information's 8 December 1975 decree renamed the erstwhile colonial broadcaster Rádio Nacional de Angola (RNA; Angolan National Radio) and gave it the call sign "from Luanda, capital of the Popular Republic of Angola, this is Angolan National Radio broadcasting in connection with the national network of transmitters."[4] It restructured the broadcaster, though actual change occurred only after the violent, messy purge and centralization of state and party power in the wake of the 27 de Maio.

The events of 27 May 1977 forced the party leadership to realize how much they needed radio expertise. In consequence, President Agostinho Neto appointed a professional from the late colonial period to the directorship. RNA became a high-functioning state institution between the late 1970s and the early 1990s (when new laws liberalized media in the lead-up to the first elections in 1992). Radio professionals, as they tell it, constructed an efficient state institution. Because of the privileged position of radio as a medium of communication needed for war and nation-building in the years between 1977 and 1992, radio employees could innovate without too much party/state interference. Privilege presupposed party loyalty. Over time, loyalty and professionalism pulled in distinct directions.

The story of RNA in this period straddles two lines of analysis in the literature on Angola in the 1980s: one emphasizes economic decline and mismanagement in the general economy and institutions of state, and the other emphasizes competence and international oriented expertise in Sonangol, the national oil company. Much of what we know of Angolan political economy in this period comes from the work of Manuel Ennes Ferreira. He charts both the ambitious plans of the new state, the budgetary and political diversions of war, and economic decline.[5] Ricardo Soares de Oliveira has written about the exceptional status of Sonangol, "an island of competence thriving in tandem with the implosion of most other Angolan state institutions."[6] Safe from centralized economic planning and the depredations of patrimonialism, Sonangol offered continuity across the rupture of independence, maintaining staff, professional training, and confidence among foreign investors.[7] RNA sat in between the protected, economically buoyant Sonangol and the failing institutions of state: it was a semiprotected, successful state institution. But even as it emitted the state frequency, it was shot through with oppositions—first party dissidence, then rebelliousness borne of a sense of expertise.

The nationalization of radio is a story of continuity: personnel from the colonial period, the use and completion of infrastructure built in the late colonial developmental/counterinsurgency phase, and the continued beating on the propaganda drum. It is also a story of change: rupture in the MPLA, novel kinds of programming, new international partnerships, and experimentation that Dibango's "Weya" announced. It is a story of institution-building under dire economic circumstances and political centralization. This is a story of how state employees negotiated professionalism (elevating technical skill and expertise) and party loyalty (the demand to follow protocol, to perform fealty in ever narrower forms). Unlike Sonangol or the factories Ennes Ferreira studied, RNA is not a revenue-producing sector of the economy, yet it offers insights into how state actors melded socialist ideals and capitalist practices to create a resilient, state/party media institution. It renews the significance of the immateriality of the state grounded in technical infrastructure and human work.

INFORMATION INDEPENDENCE AND THE TRANSITION

When Agostinho Neto declared independence for Angola in the name of the MPLA just after midnight on 11 November 1975, the MPLA already controlled the EOA. A convergence between the interests of late colonial period radio professionals, mostly white and some mixed- race individuals, nearly all MPLA supporters, and a few MPLA cadres returned from exile who had worked in information created a basis for broadcasting. RNA's current website writes the history this way:

> The Official Broadcaster, later designated as Angolan National Radio, was the first organ of the Portuguese State that, even before independence, offered a public service to the revolutionary cause. Therefore, it was the first institution whose activity made the transition from colonization to independence, constituting the most important vehicle for spreading the official discourse of the new State born in Angola. Angolan National Radio is, in these circumstances, the first Public Business of the Popular Republic of Angola.[8]

The RNA's statement elides a number of struggles. The transition was far from smooth.[9] Times were tense and arrangements related to information complex. Nor did the MPLA have a complete hold on the EOA.

Over the course of 1975, the agreements of the Alvor Accord unraveled, tensions flared, one party sought to oust another and establish strongholds in Angola's cities, and Portugal failed to enforce the agreement.[10] The Portuguese Armed Forces Movement (MFA) granted each party a thirty-minute broadcast segment.[11] Attacks between the parties characterized life in Luanda until August 1975. One attack occurred at the radio station itself. The capacity to communicate with the entire territory (even if only by rebroadcast) made the station a critical site in Luanda's newly politicized urban geography and a key communications channel for each of the three parties and the MFA.

Shortly after independence, however, life at the radio changed. The MPLA, having secured the city and the station, seemed to have more pressing issues at hand. Instead of putting someone from the ranks of guerrilla radio—say, Paulo Jorge, Mbeto Traça, or Adolfo Maria—at the helm of the Official Broadcaster (EOA), the president appointed Alexandre de Carvalho, Mateus "Mbala" Neto, and then Santos Matoso, men with no training in radio or even journalism (but all adepts from the party's Department of Information and Propaganda).[12] Not for a moment did the party consider any of the very skilled, and very loyal to the MPLA, professionals from within the radio station. Perhaps reflecting the disregard with which they were treated, radio professionals characterize this period as disorganized and lacking in discipline, a liminal time between the stifling order of the colonial period and what they would later create.

For a party that sprouted from a liberation movement that had managed information so strategically, the MPLA was not very thoughtful about radio in the early days of independence. Perhaps they were offering rewards and prioritizing loyalty or service in active guerrilla struggle, or emphasizing ideology over technical skill. Whatever the case, radio professionals and technicians, some of the few young Portuguese who supported independence and stayed, felt alienated. Until the 27 de Maio.

27 DE MAIO AND RNA

Independence in Angola marked the birth of the nation and the birth of civil war. It offered a new canvas on which to paint old schisms within the MPLA. In particular, what has come to be known as the 27 de Maio, the day a group of young dissidents in the MPLA, led by Nito Alves, former minister of the interior, are accused of having attempted a coup. Some say he plotted a coup. Others insist it was only a protest. Lara Pawson concludes in a recent work that it was likely both.[13] Jean-Michel Mabeko Tali calls it "a very ideologically motivated insurrection," and he underscores that Alves and others agreed on the need for a change in power but debated how best to execute it.[14] President Agostinho Neto called them *fraccionistas* (factionalists). The MPLA, with support from Cuban troops, stymied the insurrection and popular mobilization. A manhunt, purge, two years of imprisonments without trials, and murders followed. Estimates of fatalities ranged between twelve thousand and eighty thousand.[15] As Tali notes, the number remains impossible to calculate even today.[16]

Tali refers to this as the last of a series of historic dissidences in the movement but one that was distinct for the moment in which it occurred (i.e., with the MPLA in power), and for the long-lasting effects on Angolan political culture and MPLA rule.[17] While the MPLA is the subject of much scholarly, journalistic, and memoiristic reflection, only since 2002 has 27 de Maio received much attention. The chilling effects of the murders and purge on the party, especially hard on its young cadres so soon after independence, contoured politics and scholarship for generations. In the wake of the events, President Neto announced: "There will be no pardon." Members of DISA (the Department for Information and Security) executed the eleven people the state identified as coup plotters without trial. Among them were the leaders, Bernardo "Nito" Alves; José Van-Dunem; Sita Valles, a Portuguese communist who had joined the Angolan struggle; and Santos Matoso, then director of the Official Angolan Broadcaster (EOA).

Some saw Matoso as an ineffectual director of the radio station. Others deemed him a keen political thinker. He was one of a few who rotated through the position in the first two years of independence, signaling instability, the shuffling of posts among political appointees, and MPLA leaders' lack of clarity about the role of the radio for the party and the new nation.

Many Portuguese fled Angola at independence. While a handful of radio professionals remained (mostly those born in Angola who in the colonial period had been considered second-class whites and entered civil service jobs like those at the Official Broadcaster only with difficulty and through *cunhas*, inside contacts), some felt sidelined by the radio's new administration.[18] Most kept doing their jobs despite what they considered poor leadership and feeling marginalized. Young sound engineers and journalists put their skills to work on new programs.[19]

Emanuel "Manino" Costa was active in the Popular Neighborhood Commissions (especially his in Sambizanga), which were the site and the source of popular mobilization. Having graduated from the liceu and served in the Portuguese military, he started the radio program *Kudibangela* for the MPLA in August 1974. By August–September of 1975, the show had turned its focus to social issues: the haves and the have-nots, the challenges of class difference, the overlap of race and class—questions that could not but touch the party structure. According to Manino, a diverse group of people (Mbala Neto, Adelino "Betinho" Santos, Costa Benjamin, and Rui Malaquias) composed the show's team—various language speakers from different parts of the country—and each day of the week they featured one local language on the show.[20] It ran in the mornings for thirty minutes and was rerun in the evenings. Later it ran for an hour, until the party canceled it completely.

The emphasis on questions of race and class in a revolution that emphasized equality, headlined by a leadership heavy with mixed-race and white individuals, hit a disturbing note. Information minister João Filipe Martins said in a speech announcing the guidelines for media institutions in December 1975 that media should serve the working masses, fight bourgeois customs and habits and the addictions of colonialism and capitalism, and respond positively to the generalized popular resistance. They should fight against the divisiveness of tribalism, racism, and triumphalism. He called for greater control by the MPLA over information. Then he called out *Kudibangela* by name. It abused its popularity, he said. Broadcasting "commands" not previously approved by the MPLA, the program overstepped the bounds of its political remit.[21] This was the beginning of the end. Manino Costa remembered that President Neto soon thereafter invited the crew to dinner and told them, "Comrades, it is convenient for the MPLA that *Kudibangela* is no longer aired."[22] And that was that.

Tali described *Kudibangela* as an unabashedly Nitista program. They aired Nito Alves's speeches (he was a member of the party's Central Committee) in their entirety, just as the party's paper, *Vitória É Certa* (Victory Is Certain), under the control of the Henda Action Committee, printed them all. Not even President Neto's speeches received this much coverage.[23] Alves sympathizers controlled particular organs of information: *Diário de Luanda*, *Vitória É Certa*, and *Kudibangela*.[24] They understood the significance of communications.

Michael Wolfers and Jane Bergerol, English journalists and Neto loyalists, characterized *Kudibangela* as "a popular radio programme [that] bemused its audience with talk of differentiation between anti- and multiracism. Its young broadcasters were becoming involved in the hair-splitting of Luanda debating societies."[25] Despite their dismissive tone, the comment underscores the program's wide appeal and the engagement of its broadcasters with the political questions debated in neighborhood commissions and political committees over how to craft a postcolonial political position relative to race.[26] No small task, and not one that Neto had resolved. Indeed, Neto's continual sidelining of racial and ethnic questions across the years, a series of internal dissidences, and the party's position of ignoring the problems of colonialism to concentrate on the class struggle and the construction of the Homem Novo (New Man), generated discontent.[27] Another observer described the program to author Lara Pawson as never anti-Neto, even if it was pro-Nito.[28] Tali employs the same description to discuss Nito's position and actions throughout the period leading up to the attempted coup.[29]

Whatever disagreements exist regarding the intentions for 27 de Maio (coup or demonstration), everyone agrees that on the morning of 27 May 1977, two days after the MPLA's Central Committee expelled Nito Alves and José Van-Dunem, and several months after *Kudibangela* stopped broadcasting, people awoke to the show's familiar opening call sign: "Kudibangela, weya, weya!" and knew that something was afoot.[30] Wolfers (then a collaborator at RNA) remembers station employee Rui Malaquias and other members of the *Kudibangela* team, each accompanied by soldiers, commandeering station broadcasting "in the name of the revolution."[31] Taking the station, the Ninth Squadron (an elite military unit) barracks, and murdering key MPLA ministers made them a credible threat. Neto called on the support of Cuban troops, who, with Angolan general Onambwe, entered the radio station grounds with tanks. For a moment, Cuban voices spoke over the microphones of RNA, securing the station in the name of Neto.[32]

CENTRALIZATION IN THE WAKE OF 27 DE MAIO

The purge after the 27 de Maio crushed the young cadres of the MPLA. The period of rich and ardent debate ended in fierce posturing, abrupt structural changes in the form of party politics, and blood and torture and fear and death. *Poder popular*, "people's power," once a grassroots phenomenon of urban neighborhoods and Luanda's *musseques* became an echo chamber for slogans formulated in the high-rise of party headquarters in the central city and passed down to loyal and aspiring militants. In a stroke of irony, the party narrowed its politics but extended its acronym—adding PT (*Partido do Trabalho*/Workers' Party) to its title. It became a vanguard party that controlled the state: membership available to the few; a logic of meddling, spying, and gossip for the many.[33]

The 27 de Maio taxed internal party governance and ideological integrity. As political space narrowed, the economic situation continued to decline. Economist Manuel Ennes Ferreira's work traces the development of MPLA policy in this period. The Constitutional Law of 1975 placed the state in the position of "guiding and planning the national economy." By the end of 1977, the party took over state functions (replacing ministries with departments). The state committed to socialism and underscored the central role of the party in that process: "The year 1977 represented the definitive assumption of the socialist route and of a centralized and planned economy. . . . It was declared the 'year of the First Congress of the MPLA and of the creation of the party' (Marxism-Leninism) and the 'year of production for socialism.'"[34] Centralization took time and had to adapt.[35] The 1978 Constitutional Law, on the heels of the First MPLA Party Congress in 1977, announced central state planning of the economy.[36]

Besieged by foreign aggressors, youthful political exuberance, and critique from inside the party, the MPLA struggled to implement its socialist ideals.[37] It zipped itself up in vanguardism and focused political economic ambition. The MPLA-PT Party Congress in 1977 and the Extraordinary

Congress of the MPLA-PT Party in December 1980 shaped laws, institutions, and planning for the early republic.[38] These congresses reinforced and "legitimized" economic planning and centralization.[39] The MPLA-PT government would set itself about the business of social and economic development based on socialist (that is, state- and cooperative-owned) property, of organizing a centralized and planned economy (aimed at raising production levels to those of 1973), constructing a socialist society, and creating a Marxist-Leninist party.[40] By late 1977, that now smaller, more ideologically strident MPLA made centralized control of economic development critical to the construction of a socialist society. Though RNA was not a productive sector of the economy, these developments shaped the life of the broadcaster through party diktat regarding work organization, through the stultifying effects on the economy, and through the stresses placed on the daily lives of the Angolan population.[41]

The 27 de Maio produced profound structural changes at RNA. In early December 1975, Information Dispatch no. 2 transformed the Official Broadcaster (EOA) into Rádio Nacional de Angola (RNA) and created a monopoly on radio broadcasting controlled from Luanda.[42] But it was only after 1977 that RNA began to put this law to work. RNA integrated the radio clubs and Rádio Comercial into the national broadcaster and the state closed Rádio Ecclésia. Radio employees inventoried each former club's equipment (each had different material and distinct maintenance needs, resulting in a phased transition) and created work and station operational criteria that could be applied across the country. This restructuring created the first truly national, territory-wide network (something the EOA only dreamed of).

Announcements of the new economic policy followed the MPLA's First Party Congress held in October that year and its declaration as a Marxist-Leninist Workers' Party. RNA played a key role in educating listeners about these changes. Angolans could follow the party Congress on the radio, where it had its own program: *1° Congresso* (First Congress). The program opened with martial music then segued to the voice of a young woman and the well-formed pronouncements and funky sound work of the radio's best technicians—lively music, dramatic interludes, party youth singing, and representatives from the party's women's wing (Organização da Mulher Angolana, OMA) chanting about independence—to announce a program dedicated to the Angolan people from Cabinda to Cunene, especially members of the MPLA. The announcers noted that in its twenty-year history, the party had been unable to convene a party Congress. Obstacles such as war against the colonialists, the need for clandestine struggle, and then the factionalists had impeded normal proceedings.[43] The official newspaper and radio reminded people that the political and economic transformation they beheld resulted from the MPLA's successfully negotiating nefarious internal and external forces. *Jornal de Angola*'s director, Costa Andrade (Ndunduma), wrote what many considered incendiary editorials, often read on the radio, imploring the population to vigilance against factionalists and antirevolutionaries.[44] Radio and newspaper reinforced a new centralized information system that shored up the MPLA-scripted narration of nation.

Already in the midst of the roiling debates of the mid-'70s, the president had abolished the Ministry of Information at the MPLA's third plenary meeting in October 1976: "Information is no longer a government body but a body of the Party and therefore should reflect the Party's preoccupations."[45] All media fell under the aegis of the party's Department of Information and Propaganda by mid-1976. A year later, in June 1977, the involvement of the radio's director, Santos Matoso, with Nito Alves and the commandeering of the station's microphones by some former *Kudibangela* staff focused Neto on the radio's significance and vulnerability.

Newly aware of how radio gave the party access to the nation's ears, President Neto appointed Rui de Carvalho director of the radio station. A former sports journalist from Huambo, Angola's second city in the south, people across the territory recognized his voice. On 5 October 1977, President Neto visited the radio station and highlighted the key role the radio played in educating the Angolan people about the "political orientation of the MPLA."[46] A year later, radio employees voted to commemorate 5 October as the "Day of the Radio" (an anniversary that endures).[47] This date commemorates the role of the 27 de Maio in the restructuring of the radio. Instead of marking the date of the decree that made the national broadcaster (RNA) from the colonial one (EOA), 5 October unwittingly folds the history of the 27 de Maio into the history of national broadcasting (see fig. 4.1).

Carvalho reintegrated those radio technicians and journalists who had felt sidelined. He drew in others who had worked in the radio clubs. They sketched organograms, wrote lists of job functions, and homogenized the organization of programming in Luanda and at the other stations. This crew had to identify local skill and find and train new recruits. Men like José Patricio and Aldemiro Vaz de Conceição—smart, young party militants—worked at RNA before ascending the party hierarchy to the president's office.[48] The ability to communicate—to think fast, write well, and speak with conviction—proved a valuable and transferable skill.

Nation-building and authoritarianism drove this push for uniformity and conformity at RNA. The civil war pressured state ideology and policy to impose unity, while intolerance of dissent within the ranks of the party

FIGURE 4.1 The fifth of October has been designated "Day of the Radio." This day commemorates the beginning of professional, organized state broadcasting after independence. Submerged in this celebration is the attempted coup of 27 May 1977. Photo from the personal collection of Guilherme Mogas (the tall, blond man in the right of the photograph) and used with his permission.

energized conformity. Here I take a cue from Anne Pitcher, who argues that in the modernist rationalization of agriculture through communal villages and centralized production strategies in Mozambique, "conformity also served nationalist goals: making everything the same helped achieve national unity."[49] Pitcher argues that the centralized and planned economy in Mozambique, like that in Angola, knit together projects with competing interests: socialism, modernism, and nationalism.[50] In Angola, the intense internal struggle to enforce unity within the party produced a tremendously strident political discourse, pulled taut across a series of contradictions. Rádio Nacional de Angola bore this out in particular and peculiar ways.

POLITICS VERSUS PROFESSIONALISM

President Neto joked on his visit to RNA that the radio's installations were "a palace compared to the cabana" of *Jornal de Angola*, the state daily.[51] The radio gave the party/state the physical presence and technical reach to both look and sound like a state. Even in an unfinished building, the radio station and its programming projected the party's/state's ambitions. At independence, the "Plan" (PRA) of the late colonial period had not been completed—the auditorium had yet to be built, the sound studios were unfinished, and the five-story building of administrative offices slated for construction existed only on Le Corbusier–trained, Angolan architect Carvalho's giant sheets of drafting paper. Fernando Alves, the famed voice and programming genius of Rádio Comercial, having fled the South African invasion in Huíla in 1976, lived temporarily in one of the studios.[52] After the shocks of the 27 de Maio settled down, a can-do, building-the-new-nation, in-it-together spirit pervaded the place.[53] Still, ambitions, desires, and team spirit needed technical power and programming expertise. Otherwise, RNA would just be an echoey, brutalist edifice.

Emerging as a state embattled in a civil war entangled with the tentacles of global superpowers and regional warlord apartheid South Africa, propaganda was a priority. In the propaganda effort of the anticolonial struggle and now in the civil war, radio held a privileged place among other media.[54] It could reach places in the national territory occupied by UNITA rebels and South African invaders. As the war endured and destroyed transport infrastructure, this medium of communication became even more important. Radio was a technopolitical necessity. It helped forge national unity in the context of civil war and the global Cold War.[55] This brought professional expertise to the fore.

For the MPLA, the radio was a key mass media form (media da diffusão massiva, MDM).[56] Of the four key media: newspaper, radio, news service, and television, radio had the largest number of employees (1,688 in 1990, for example) and the biggest budget.[57] *Jornal de Angola*, in the old haunts of *Diário de Luanda* (Luanda Daily), was the state paper. But RNA had a much greater range in a country with a vibrant oral culture, high illiteracy rates, and low population density spread out across a large territory. Angola boasted the prime conditions for the problematic Jeffrey Herbst calls "broadcasting power," though he's not concerned with radio communications per se.[58] Like the postcolonial states Herbst describes, the independent Angolan state extended infrastructure throughout the national territory, surpassing the ambitions of the colonial state. Radio, more than newspaper and more quickly than television, offered the perfect technology. Fighting destroyed transport infrastructure like roads throughout the 1980s. Radio broadcasting provided an immaterial sense of connection that overrode the breakdown that characterized material realities.[59] The single network could unify the country through the ether, sending the sound of the state to the reaches of the territory it did not control physically. Contra James Scott and Jeffrey Herbst, the state used an immaterial form (words broadcast via sound waves) to project its power in place of the material organization of space and structure. State power relied less on the material than on the immaterial frequency to manifest a consistent presence.

If politics demanded the technology, radio professionals gave it life and filled it with meaning. Unlike Sonangol, which experienced a smooth transition from the colonial period to independence, the transition period disrupted work at RNA.[60] But like Sonangol, once Neto appointed a new leadership, they received the latitude to do what was necessary. The characteristics that Soares de Oliveira identified in Sonangol employees held true for the RNA leadership: their loyalty was unquestioned, they were respected technocrats, and political support for the project protected them.[61] Official MPLA Party organs, with the approval of the Political Bureau, nominated directors and subdirectors of sectors (head of editing, for example).[62] Working with the party's imprimatur gave radio leadership a degree of maneuver. It protected the RNA leadership against criticisms from ministries, provincial commissioners, and other institutional representatives.[63] This made it possible to get things done.

The first directors after independence highlighted their technical skill and expertise and pointed out that the directors of the transition period lacked these qualities. For the RNA directors and employees working at RNA in the years 1977 to 1992, this shift from ideologues to experts is a throughline in the story they construct of a high-functioning state/party apparatus.[64] In their telling, unlike other state institutions, the expertise of radio employees manifested in the RNA's ability to deliver its annual reports on time, make do with a limited dual-currency budget (dollars for technical equipment, kwanzas for salaries, maintenance, etc.), and broadcast high-quality programming even under trying economic, technical, and political conditions (civil war, electricity and water cuts, international territorial invasions, support of neighboring liberation movements, etc.).[65]

The demands of creating a well-run, efficient station (with full programming in a variety of languages, shows that ran on time, news flashes about the war and other breaking stories, coordinated from Luanda but in conjunction with ten provincial stations) often conflicted with the ideological imperatives the party imposed on radio business. A note on the radio's history in a binder in RNA's documentation center, a mini-library for journalists in the pre-internet era, is telling. Written by a consultant for the Ministry of Information at its reconstitution in the mid-1980s, it reports that many journalists interpreted the closure of the Ministry of Information in 1976 as a threat: "The Party's decision to opt for the politico-ideological education of journalists, in detriment to their technical-professional education, was a limiting factor in the development of National Information, due to the absence of a coherent policy of education for creating technically and scientifically prepared personnel."[66] The author underscored that technical and scientific training are the base of radio expertise. Despite the highly charged political atmosphere in newly independent Angola (or perhaps because of it), and the requirement of MPLA party support if not card-carrying militancy to work at the radio station, many RNA employees saw themselves first and foremost as radio professionals. However, other journalists and technicians remembered their political commitments as on par with the professional ones, or hoped that quality work (polished technically, tight professionally) might embody the politics.[67] With few employment options and with so few employees, journalists and radio professionals kept working. The radio leadership made recruiting and training new employees a priority.

Due to the flight of skilled Portuguese at independence and the dismal educational system provided in the late colonial period, Angola did not have the skilled professionals and workers it needed to develop its economy. MPLA discourse emphasized this lack of trained employees. The fallout from the 27 de Maio exacerbated the situation. Thousands of young people were killed. Many left the country or failed to find appropriate work or even the motivation to do it; some were disqualified from state work by their political dissidence.[68]

The MPLA prioritized ideological education in schools and ideological training in workplaces. Policies on work explained that "training should permit [Angola] to obtain 'politically, scientifically, and technically capable workers' (note the order of presentation)."[69] The party anchored professional and general education in political education. Advancement in the civil service hinged on ideological alignment and party loyalty.[70] In a country already suffering from low levels of literacy and a distorted colonial educational system, now hobbled by war, this had long-lasting implications. It had devastating effects on public education, teacher training, and worker training in general.

At independence, the flight of settlers to Portugal (and to other white-dominated countries in southern Africa) reduced the numbers of skilled radio broadcasters and functioning radio stations. One of the first private businesses the state confiscated was Rádio Clube de Angola. By taking RC de Angola, the party/state announced its interest in information, and its attempt to quickly remedy what it identified as a critical problem: lack of trained personnel.[71] Taken over in late November 1975, Rádio Clube de

Angola became an education center for journalists. By 1979, the directorate of RNA and the MPLA had developed it into a more formalized and technically structured program: Rádio Escola (Radio School).[72] Courses ran from two months to two years. Angolan journalists and broadcasters taught courses in collaboration with East German radio professionals and journalists.[73] RNA also sent employees abroad to study in East Germany, Cuba, Czechoslovakia, the Soviet Union, Spain, and Yugoslavia.[74] Under the ideological shingle of the MPLA, and the Eastern Bloc, courses emphasized the technical and general education that made a radio journalist a professional.

Internal RNA documents from seminars, held mostly during the 1980s, emphasize the technical, historical, and general cultural education required for work at the radio station. Translated Cuban pedagogical documents underscore the basics of news writing and the sociopolitical role of the press in a revolutionary society.[75] Dozens of pages of copied documents on locution of the news, with drawings of the human torso, discussions of voice projection, breath control, and strategies to avert fatigue, circulated among radio journalists in RNA seminars.[76] One piece of excerpted material came from a Spanish National Institute of Radio and Television manual and another from Radio France International.[77] By the late 1980s, technical and professional models came from locations other than Eastern Bloc countries.

This was true earlier too. Information in copied form, often unattributed, regularly passed hands, at least among the directors of RNA. A twenty-five-page document on "Leadership," dated 1979, reads like a section of a business school textbook or a managerial primer. How to organize your time, arrange your work, use praise with employees, engage in constructive criticism, solve problems, and a list of what not to do offered management strategies for the novice.[78]

Other RNA internal documents (e.g., a collection of monthly reports from the Department of Information in 1980) dramatize the conflicts between state imperatives and running an efficient communications business with a staff overwhelmingly still in training. Five years after independence, and three years after the 27 de Maio, 1980 was still early days. The programming grid had been set at the end of 1978/beginning of 1979. The ink had only just dried on program plans, department structures, the organograms that sketched the flow of power, and the function guides that defined who did what in relationship to whom.

The RNA fell under DEPPI (Department of Information and Propaganda), a body of the ruling party, with a representative on the Central Committee, the second-highest body in the party hierarchy. In the Department of Information reports from 1980, the department, headed by famed radio journalist Francisco "Xico" Simons, found itself beset by a number of difficulties in relating to the party/state: poor communication from the president's office, lack of information from government ministries, and new directives that pummeled the morale of journalists in the regional stations (one result of centralization). The June 1980 report directs a word specifically to the RNA's director general and to the director of programs, asking them to communicate with the head of DEPPI. It notes that accompanying and reporting on the president's visits outside the country (and even to the interior) requires tremendous technical coordination (with international satellites), decisions about equipment, and program planning, but "it happens that, lately, DEPPI, when requesting a reporting crew, does not furnish any elements that can help us plan in the preparation of reporting."[79] A complaint about transport promised by DEPPI for reporting followed: "On a recent trip between Nambuangongo and Luanda, the crew found itself seated in a Land Rover between sacks of cassava, sweet potatoes, and bundles of bananas."[80] Like riding in open pickups (as sometimes happened), this endangered the longevity of technical equipment, they insisted.[81] Technical knowledge, and the protection of equipment, should take precedence over political expediency. Here we see the technical and professional concerns of journalists interfering with the attempts of the party to instrumentalize the radio for their propaganda needs.

THE SOUND OF THE POPULAR REPUBLIC OF ANGOLA

When Angolans tuned in to RNA on the morning of 27 May 1977 and heard the once familiar "Kudibangela, weya, weya!" the sound of this banned program inspired skepticism, even worry. Silenced six months earlier, it struck a dissonant note.[82] With Rui de Carvalho quickly installed and a programming grid set by early 1978, broadcasting settled down and radio sound stabilized. In the semiprotected halls of post–27 de Maio RNA, broadcasters crafted distinct and memorable programs. Angolan listeners and broadcasters remember snatches of jingles and shows, drawing the sound of the past through the filter of memory and utterance. Programs were diverse, even if they toed the political line. Despite the fact that RNA was the only station with territory-wide broadcasting (or would eventually be), it never took its listenership for granted.

The directorship appointed by President Neto after the 27 de Maio acted quickly to professionalize RNA and its sound. Infrastructure took time and money to build. Novel programming relied on creativity and will. At independence, the RNA continued to sound official. This was not surprising

FIGURE 4.2 *Piô Piô* was a monthly show for children and adults. Elisabete Pereira traveled briefly to Cuba after independence as a media professional. She was inspired by the excellent Cuban programming for children and started a daily children's program and monthly live show. Photo from the personal collection of Guilherme Mogas and used with his permission.

given that the radio professionals who stayed on had been part of EOA. In the late colonial period, Ecclésia sounded more Angolan than the EOA.[83] It took training at Rádio Escola, and the entrance of new employees (some from Ecclésia after it was closed), for RNA to Angolanize in sound and style.[84]

A number of programs established in this period, under the leadership of Rui de Carvalho, himself a journalist, continue to this day: *Azimute* (a program on the economy); *Antológia* (a program on Angolan literature); *Bom Dia Angola*; *Boa Noite Angola*; *Re-encontro com Àfrica*; and *Manhã de Domingo*. Others had shorter but memorable runs: *Trabalho e Luta*; *Sala Piô* (a children's show with its monthly event, *Piô Piô*); *Cassulinhas de Bola*; and *O Crime Não Compensa* (see fig. 4.2).

The programs that endured structured the quotidian and weekly soundscape through the ideological exhortations of the party and the Pan-African cosmopolitanism of RNA's sound technicians. *Dia Novo* (New Day) filled the air of the very early morning hours, followed by *Good Morning Angola* with standard morning fare, and *Good Night Angola*, broadcast at 10 p.m. Openings such as "Good night, Angola! So that tomorrow we can have a good day of work! . . . A program truly heard from Cabinda to Cunene," contoured the night in socialist expectation and nationalist territorial ambition.[85] It proclaimed national listening space with the MPLA's slogan "from Cabinda [in Angola's far north] to Cunene [in Angola's far south]!" and declared the day for labor activity. Aimed "to educate, inform, and entertain," *Boa Noite Angola* previewed the RNA's sonic offerings for the week.

Newscasts at 10 a.m., 1 p.m., and 8 p.m. punctuated the workday with the explosions of war, the movements of diplomacy, the thunder of national mobilization, and the roar of international solidarity (see chapter 5 for examples). The 1 p.m. news often found families around the lunch table listening to reports about South African bombings of Lubango, UNITA attacking water conduits, or antiapartheid war crimes rapporteurs addressing the United Nations. The movement of government forces (FAPLA; People's Armed Forces of the Liberation of Angola [MPLA]), described as a "strategic retreat," optimized defeat in euphemism.[86] The news celebrated defeats of UNITA and imperialist South African troops, while martial tones and heroic timbres structured the remade, independence-era *Angola Combatente*. Quotidian comings and goings wove the bass structure of the public agenda, where citizens communicated the death of friends and loved ones, or the safe arrival of someone to the capital or to a provincial town. It linked the rhythm of the life cycle, daily arrivals and departures, of small towns and big cities across the vast expanse of the Angolan territory.[87]

Based in Luanda, RNA employees created a number of programs that used local, live programming to bridge the studio to urban life. *Cassulinhas de Bola* (Soccer Kids) was a weekend program started by Ladislau "Lau" Silva that ran for eighteen months. *Cassulinhas* made things happen. The program staff organized thirty children's soccer teams in Luanda for kids between the ages of six and twelve. They arranged team sponsorship from national companies and ministries (the national soft drinks company and the navy, for example). These parastatal sponsors provided transportation to the games and a snack for the teams (no small challenge in the face of food shortages). Players and referees from the national team served as coaches and arbiters, creating an inspiring pedagogical environment. All the players received sports shoes produced by the state company Makambira and shorts, jerseys, and socks from the state clothing company Confex. Reporters related the games live. Lau Silva recalled that this provided a healthy and fun physical activity for children in a period otherwise heavy with crisis and news of war and food shortages: "We needed to counter-balance that, we needed freedom from it."[88] Eventually provincial broadcasters organized their own *Cassulinhas*, taking the program outside Luanda.

Widely applauded, the program lasted only a year and a half, folding under the pressure of failing factories, the economic crisis, and the aggravating factor of war.[89] Other shows, like *Trabalho e Luta* (Work and Struggle) hit the streets to witness the world of work and economic production. A morning show, reporters appeared at factories to chronicle worker productivity and factory-based union activities. In offices of notary publics and ministries they recorded civil servant foot-dragging or absenteeism and alerted higher authorities.[90] "Counter-revolutionaries"—those inside state institutions (factories, ministries, state bakeries, schools), in the party, or in the neighborhood, along with those outside (UNITA rebels, imperialist forces), required constant vigilance.

RNA sent its microphones to amplify stories of corruption and antirevolutionary actions. This dynamic led to some of the most interesting and problematic reporting and programming for the party/state. Turning microphones to counterrevolutionary activity might focus on individual failure (the incomplete erasure of bourgeois habits), but it could also increase the volume of state/party failures to deliver services and training.

The popular radio detective show *O Crime Não Compensa* (Crime Doesn't Pay) walked this line. Investigating cases of counterrevolutionary activity, detectives Goodnight and Spitfire visited different cities and workplaces throughout Angola. The one episode for which I could find a partial recording found them in Benguela. In that central-southern city, a young man offered to sell them spray deodorant, a black-market product unavailable on state store shelves, either by the can or by the spritz. Lau Silva, who performed Agent Goodnight, said the program confronted quotidian problems with humor and a dose of political, social, and economic criticism. This made it popular with listeners. A part of the early programming schedule in the years 1978–80, *O Crime Não Compensa* fell victim to the problems it ridiculed. When worker indiscipline sounded too much like institutional failure, the MPLA suspended the show.[91]

RNA played a critical role in the production of music after independence. Sound technician Artur Neves and journalist Artur Arriscado traveled the country recording the musical traditions of the different provinces. Once the recording studio CT1 opened, RNA functioned as the only, and then the primary, producer of music in independent Angola.[92] The 27 de Maio also reverberated through Angolan music. Three of the most popular and high-producing musicians of the late colonial period—Urbano de Castro, David Zé, and Artur Nunes—were killed in the purge. Deeply associated with Luanda's musseque neighborhoods (particularly Sambizanga, considered a base of Nitistas), their many albums became *musica* non grata. Deep, intentional scratches in vinyl discs produced in the early 1970s that remained part of RNA's music library in the late 1990s testify to the sonic censorship at work.

RNA recorded revolutionary songs that promoted yearly MPLA slogans and praised fallen heroes. It was also the home of the Merengues, the touring band of Angolan sounds, including semba, headed by Carlitos Vieira Dias. And it was the launch pad for the group S.O.S., started by young employees of RNA (Eduardo Paim, Bruno Lara, Nelson do Nascimento, Bruno Levy, and with songs written by Carlos Ferreira). S.O.S. generated a new sound and genre for a distinct moment.[93]

RNA organized public music events, including a twenty-four-hour music festival on Luanda's island in 1981. This caused a shock within the party: it suspended three of the radio's directors briefly. Programs such as *Os Tops dos Mais Queridos* (The Top of the Most Loved List) solicited audience participation and promoted local sounds. And the *Top dos Cinco* (a Top 5 for the PALOPS) was broadcast from the Cidadela sports stadium. CT1 did not have the equipment to do a show in a stadium of that size. Bruno Lara and one other technician traveled to the United Kingdom where they rented equipment from the company that did sound for Pink Floyd. They wanted to do something different. They hired a parachutist and reported his jump into the stadium; he landed in the goal zone, carrying broadcast equipment in place of an emergency parachute.[94]

Music structured programming. Documents spoke to the role of music in society in general, the history of music theory and musical genres in the West, and its importance in creating a sound for programs and the rhythm of the broadcast day. The thinking was that upbeat music should open the day, motivate people for work and school between the hours of 7 and 9 a.m.; and it should greet them as they arrived home in the afternoon and early evening. Listening music swathed the nighttime in reflective sound. Angolan music interspersed all music and other shows, a consistent part of the national soundscape, not just a specific part of the day.[95] The sound of RNA, the sound of the People's Republic of Angola on the airwaves, struck various tones: celebratory for party organizations (those of women, youth, and workers); informed for programs on hygiene, agriculture, and education; patriotic when mobilizing for party events, national production targets, and vigilance against imperialist aggression; and cheerful for the children's programs.

NATIONAL INFRASTRUCTURE — BUILDING THE NETWORK

Given his excellent reputation as a manager, President José Eduardo dos Santos sent Rui de Carvalho to run the television station (still broadcasting in black and white in 1982). In his place the president nominated Guilherme Mogas. Mogas, unlike Carvalho, was not a familiar radio voice or a seasoned journalist. He was a highly skilled technician, the head of the infrastructures department, and someone who had traveled regularly to all the provinces. As Carvalho said, he possessed "a national vision."[96]

The civil war, a struggle over the nation, unfolded in a regional space, traversed by international interests. The MPLA's involvement in the war of words around Namibian independence and in defense of its own territorial integrity (the subject of chapter 5), put a priority on Mogas's capacity to extend the FM network and to connect provincial stations in real time to the Luanda studios. The ideological structure of the party and organizational hierarchy imposed by monopoly subordinated provincial stations to the central studio in the capital, but Mogas set out to create easier communication.[97] Mogas's mastery of technology could realize political plans. President dos Santos, trained as an engineer, saw Mogas as the man to advance the technopolitical project of the radio.[98]

At independence, RNA inherited the most advanced Portuguese radio station, but one that lagged behind the ambitions of the MPLA for national broadcasting. The technopolitical plans of one state did not map onto those of the next. Tied into an imperial communications system, oriented to Lisbon more than Luanda, the EOA did not participate in regional broadcasting networks. Precarious technical conditions troubled good reporting. Reginaldo Silva, then working as a reporter, accompanied President Neto on his first foreign visit, to the neighboring Republic of the Congo in early 1976.[99] He and reporter Pedro Almeida flew to Brazzaville. There they went to the radio station (the same strong state broadcaster that had broadcast *Angola Combatente*), to request that the station cede them airtime. RNA tuned in, recorded their report, and rebroadcast it to the Angolan territory.[100] Lau Silva remembers reporting on the MPLA's first meeting outside Luanda (in Kwanza Sul Province) at the Kaddé coffee farm.[101] RNA sent Silva and two technicians: one Angolan and one Cuban. The Cuban technical support officer had to find the highest hill or the tallest tree in which to situate himself with a broadcast antenna and 600 meters of cable to reach the reporter. These stories amplify the limits of the colonial broadcasting system and the hurdles that new states faced in converting institutions to different political purposes. No matter how state-of-the-art, the station was not easily or instantly converted to a national broadcaster.

The conditions of life after independence—economic crisis, the flight of Portuguese technicians and bourgeoisie, a centralized economy—created their own constraints. Indeed, RNA required a series of departments that it never needed in the colonial period for day-to-day functioning: transport and mechanics; carpentry and metalworkers (to build furniture and do construction work); and a maintenance department with expanded capacities, to fix, for example, broken typewriters. To convert RC Angola to Rádio Escola required creating a mini construction business inside RNA in order to do the buildout. To finish off the auditorium and CT1 (the recording studio), Mogas and Domingos Neves had to learn about new materials and sonic phenomenon such as wall sound resonance. While under Carvalho's leadership, Mogas opened the Department of Technical Material Supply, an independence-era novelty and necessity in the face of economic crisis and empty store shelves.[102]

Under Mogas's leadership, a national network became a reality. The first step was twenty-four-hour broadcasting. This was a technical matter—done to preserve the transmitters. Under Carvalho, RNA broadcast from 6 a.m. until 1 a.m. and then shut down for five hours. Mogas knew this threatened the longevity of the transmitters. He instituted around-the-clock broadcasting almost immediately after his appointment. Then, under his leadership, RNA created a national network that could broadcast from any province in the country through Luanda in real time. Carvalho had initiated the nationalization of radio and built infrastructure.[103] But news reports from the provinces arrived two to three days after events occurred, on recorded reels placed in aluminum suitcases via a dedicated mail service that traveled overland and by air (yet another department and internal infrastructure novel to this period and these conditions).[104] By the time of Mogas's appointment in the early 1980s, this system was outdated. Journalists in the provinces complained vociferously about the situation. So they built a cable to connect the Luanda studios to the downtown central telephone service, from which they could access the other stations by phone. The cable was made locally with the help of an engineer, another feat no one imagined possible but a necessity when imports were rare in the early 1980s.[105]

Other operations brought services to radio employees. The radio's directorate—composed of the heads of all departments—strove to produce top-of-the-line programming. They organized seminars on locution, on writing news flashes, and on reporting. The MPLA demanded their attendance at the Escola do Partido (Party School); their poor attendance sometimes got them in trouble.[106] Party directives dictated salaries for employees, who were civil servants. So the radio's directorate worked to create "indirect salaries" and "indirect increases" to keep their workers happy. Among them: a cafeteria with partially subsidized meals, soccer fields and a set of intramural teams, child care for employees' children, a health clinic, bread, two crates of beer per month, and cooking gas (requested through the ministries and distributed via the radio station). All of this neatly realized socialist vision of happy and healthy workers exceeded the organization of the workday and hierarchy (workers meetings, union bases within the workplace, and regular employee meetings with radio leadership) that party directives mandated.[107]

In addition to offering food and proper working conditions for RNA employees, the radio directorate wanted to maintain a clean, modernist space in the mid-century building. The child care area had a similar justification. The directorate would not tolerate female cleaning crews doing their jobs with babies on their backs or cooking meals on small charcoal burners in the marble halls or the interior gardens, as had begun to happen. Mogas put it this way: "We needed to attack the cause, not the consequences."[108] For him, this was a technical, not a cultural, question. The result was a cafeteria and a day care, sparkling marble floors, and brilliant glass windows.

The RNA leadership relied on secret, private enterprise to realize modernist, socialist ends. The cafeteria is one example. Capitalist enterprise existed even at the height of Angola's socialist republic and in institutions of the state. What had formerly been one province had been sliced in two, and the new province needed a new capital at Lucapa. The state imported a number of prefabricated materials but then abandoned them, RNA-Luanda learned from the provincial station. They sent a truck and hauled a ready-made cafeteria back to Luanda—complete with forks and knives and pots and pans, but no assembly instructions. It took months to figure out how the puzzle pieces fit together, but radio employees who had little disposable income and a bit of free time took it on as a challenge. Once built, the cafeteria sold tickets for meals at half the total cost. Workers left half a ticket stub in the cafeteria box and the radio management then paid the cafeteria manager (a private manager hired without party knowledge) the difference for the number of people served. RNA's directorate dined in the cafeteria alongside other radio employees to maintain quality control of the food and promote morale.[109]

In the midst of an economic production crisis, provisioning the cafeteria with meat and produce for hot meals presented its own challenge. Management improvised a system of barter: state goods and services for private produce. Mogas convinced a Lubango-based farmer whose farm had ground to a standstill to produce for RNA. He hung the RNA logo on the gates at the property's entrance, though nothing about the ownership of his farm changed—a boon in a period awash in nationalization and confiscation of private property.[110] RNA provided him with tractors and transport for the farm, and in exchange RNA received a certain tonnage of produce and meat per month. The national airline's (TAAG) charter subsidiary regularly flew merchandise that arrived in Luanda's port down to Lubango and would return with the produce. For its services, TAAG received half the produce and RNA the other half. Both parastatals thereby provided at least one substantial meal a day for their workers.

The MPLA turned a blind eye to all of this. Because the radio station produced quality programming and entertainment (musical shows, music competitions, and soccer tournaments for children), and because the radio was a model of organization (at least in their telling), no one complained.[111] As Mogas tells it, two forces created national unity in the 1980s. One was the war—soldiers from Cabinda were sent to fight in Kuando Kubango, for example. The second force was the radio. For the first time, the entire country was connected through the airwaves broadcast from a nationwide infrastructure. News of war or presidential travels or party congresses could be broadcast from Luanda across the entire territory.

After a shaky start, RNA employees built a national network that served the party and sometimes nettled it, either privately in department meetings or publicly on the air. Enlivening party propaganda (finding counterrevolutionaries meant criticizing the day-to-day operations of factories, ministries, and state offices) often threatened the party's authority. Maybe the greatest success lay in creating internal structures that made real the socialist aspirations of the MPLA and its members in the cafeteria, the day care center, and on the soccer teams. If internal administration, politics, and economic development created constant ambivalence, the regional war and its impact on Angolans and the Angolan territory were clear-cut acts of imperial aggression. Propaganda around Namibia's independence and South African invasions voiced a clarion call for action. The MPLA and radio broadcasters revived their international game on the airwaves. Where they had once shamed the Portuguese with UN declarations to help them make claims to sovereignty, they now deployed this strategy in the struggle over Namibia and against apartheid.[112]

5 "Angola: The Firm Trench of the Revolution in Africa!"

Our Anti-Imperialism, Your Cold War, 1975–92

Foi no Cuando Cubango/Cuito Cuanavale/que escrevemos a história/do mundo afinal (It was in Cuando Cubango/Cuito Cuanavale/that we wrote the history/of the world after all)

—Paulo Flores, "Semba pra Luanda," from the album *Bolo de Aniversário* (2016)

AT NELSON Mandela's 2013 memorial service, the emcee Baleka Mbete (Speaker of the South African Parliament) introduced Cuban president Raúl Castro with the following words: "Comrades, we will now receive an address from a tiny island; an island of people who liberated us; who fought for our liberation in the battle of Cuito Cuanavale. The island of Cuba."[1] With verbal economy, Mbete noted one of the key, expansive effects of Cuban intervention in Angola, a place where the Cold War went hot. Technopolitics sounded a note of continuity across the rupture of independence; geopolitics clanged with renewed significance. The MPLA proclaimed independence on 11 November 1975, having secured Luanda with the help of Cuban soldiers that held off FNLA and South African troops to the capital's south. Actors in this drama hailed from far and wide and from close to home too: Katangese gendarmes, US and British mercenaries, the FNLA and UNITA (both supported by the CIA), South African troops, Cuban soldiers, and Soviet military advisers.[2]

If Baleka Mbete made a big claim for a small island, Angolan musician Paulo Flores's "Semba pra Luanda," from 2016, is similarly bold. He sings Angola back into world historical significance: "It was in Cuando Cubando/Cuito Cuanavale/that we wrote the history/of the world after all." In this 2016 *semba* (urban popular Angolan music), Flores, a child of the 1980s and 1990s, sonically remembers this battle and contextualizes it in the world of song, a world of cultural producers. "Semba pra Luanda," with its complex polyrhythmic structure, evokes Angola's postindependence civil war and its Cold War context. Lilting and danceable, with only a few lines directly about the war, the music catalogs important musicians and carnival groups from Angola's postindependence period. The sound buoys hopes about regional liberation and solidarity to counterpose the sacrifices of the period: food shortages, losing children to the war, and bombings of Angola's south. With these lines, Flores intones pride that Angola's otherwise destructive civil war harnessed high ideals. Ideals that had a positive impact in the region at a moment when the whole world was watching. Penned in 2016, this is an assertion about the collectivist and public-minded values and gains of the Popular Republic of Angola against the post-2002 oil boom republic's individualist ethic.[3] These lines of "Semba pra Luanda" recall the revolutionary socialism that emanated from RNA, among other places, from 1975 until the late 1980s. The song evokes how radio's work imprinted minds and music and how propaganda intersected with people's convictions. Mbete's introduction echoes this as official discourse and nostalgic remembrance.

Mbete's introduction of Cuban president Raúl Castro and Flores's lines about the battle of Cuito Cuanavale's world historical significance underscore how the Cold War, the tension and hostility between the Soviet Union and the United States and its NATO allies, took on a regional and a "hot" character in Southern Africa. They notify us of a shared experience that played out in particular ways in the region's countries. Mbete and Flores inadvertently summon the ongoing debate over the history of the battle of Cuito Cuanavale, enlivening the discussion of scholars and soldiers.[4] In southern Africa, the Cold War took on a specific character: hot, knotted into post–World War II decolonization, and entwined with the racialized politics of white settler colonies in the region. Sue Onslow encapsulates this in the subtitle of her edited volume on the Cold War in southern Africa: *White Power, Black Liberation*.[5] Despite Onslow's observation that "the Angolan theatre is a recurrent reference point for the chapters in this volume, and its geographical place in the international militarization of the regional liberation struggle," writing this part of Angola's history has only just begun.[6]

This chapter offers a particular contribution to that literature by highlighting the medium of radio, its affective impact, and how propaganda works as narrative. Mbete's introduction and Flores's lyrics retrospectively underscore the agency of small countries—Cuba and Angola—in the war's big event. RNA broadcasts highlighted how Angola's civil war meshed with the regional and global battles and kept the stakes of that struggle present. Radio news announcers, under the guidance of the MPLA Party/state, used international law, Pan-Africanist ideas, and socialist vocabularies to frame the war.

RNA reported regularly on the struggle for Namibian independence and on South African aggression against Angola. RNA's sound archives hold binder after binder full of these news reports. Analyzing these reports and material from RNA, I discovered how RNA broadcasts on Namibia, as propaganda, advocated a particular ideology. Jeffrey Herf defines *ideology* as "a set of propositions that serve to interpret and lend meaning to the course of events."[7] Looking at propaganda in Nazi Germany, he argues that understanding propaganda as storytelling allows him to take the convictions of propagandists seriously while investigating the power of narrative frameworks to shape experience. "For the historian," Herf asserts, "interest in the narrative dimensions of propaganda draws attention to the connective pattern and the script that allows propagandists to assimilate events to an overall framework."[8] This is a scholarly version of chapter 3's epigraph: lying in the abstract—the framework—recasts the specifics, or the things that are true in the particular. Herf locates a "truth" in the lie and how it is effective. In postindependence, civil war–torn Angola, broadcast rhythms structured the narrative about Angola and the region in the Cold War and pieced it out daily.[9] Broadcasts consistently recounted the story of war in Angola—providing the connective pattern—across the days, weeks, months, and years. This radio propaganda provoked boredom, hope, and mobilization, the "affective rhythms" Jo Tacchi discusses.[10] Those rhythms and stories left a cultural and political residue.

The MPLA developed a new propaganda strategy after independence. That strategy emerged from a specific mix: a big event (the capture and trial of US and British mercenaries), socialist ideology, and the exchange of strategies with other similarly embattled new states. Big events—independence, battles, trials, bombings—caught local and global attention. They stick in memory, as Flores's "Semba pra Luanda" and Mbete's introduction remind us. But it was the daily broadcasts, the forging of a new soundscape through RNA news flashes across the day, that produced a distinct story.

I start with a 1982 broadcast from Huambo, in central Angola, where President José Eduardo dos Santos made a speech about South African invasions. The argument he makes and his exhortations are typical of the broadcasts of the period. Though they resound with socialist language, they speak from a very specific place. Angolans might have tired of hearing such broadcasts repeated throughout the day, but the story also resonated with them and across later decades. Radio kept the story of South African invasions, Namibia's struggle, and Angola's sacrifice sonically present.

In 1982, President José Eduardo dos Santos held a press conference in the central Angolan city of Huambo. Helicopters buzzed overhead. Microphone volume jumped and fell. The wind mauled clear delivery. Broadcast at 1300h, 2100h, and 2300h, radio listeners knew immediately he was not speaking from the sound-sealed studios of Rádio Nacional de Angola in Luanda. Of his ten-minute, forty-one-second speech, over half of it was dedicated to the question of Namibia and an associated cluster of issues: Cuban forces in Angola, South African complaints regarding those forces, the South African failure to comply with UN resolutions, and the South African invasion of Angolan territory. Hailing his local audience, President dos Santos cast these regional concerns as national questions and exhorted average citizens to mobilize for the task of defense.

Dos Santos also addressed an international audience:

> When there are no longer logical arguments against Namibia, South Africa, with support from the United States and others in its Western Contact Group, decided to insert itself into Angola's internal issues, using the presence of Cuba's internationalist forces in Angola as an argument. But they are here at the request of our sovereign government and in conformity with article 51 of the UN charter on the legitimate defense of member countries that are being attacked or invaded.[11]

He proceeded to overturn South Africa's claim that Cuba was a foreign force and to accuse South Africa of rampant aggression against Mozambique, Zimbabwe, Zambia, Lesotho, and Angola. He condemned South Africa's invasion of Angola's southernmost Cunene Province: "There are foreign South African forces of illegal occupation in Namibia and in a sovereign part of Angola. It is unjust and inadmissible that the aggressors want to present themselves as victims."[12]

Amid economic crisis, food shortages, civil war, and just three years after the end of the violent repression of the 27 de Maio (discussed in chapter 4), the MPLA party/state expended energy, money, military force, and ideological rigor on Namibian independence. The MPLA-PT declared independence an incomplete project. In radio broadcasts, press releases, and interviews, state representatives framed Angola's support of Namibian and South African independence (and specifically of SWAPO and the ANC) as part of aborted decolonization in southern Africa, stalled by imperialist powers in the West and the apartheid state to the south. Angolan state propaganda insisted that full Angolan independence hinged on Namibia's freedom and rolling back South African aggression.

Opening radio broadcasts with the call sign "Here, Angola the firm trench of the revolution in Africa! From Luanda this is Angolan National Radio broadcasting," RNA located Angola as the site of revolution on the continent and RNA as its voice (see fig. 5.1). Regional struggles were rent from imperial cloth and could not be folded into a crisp bipolar, Cold War narrative, even as Soviet military advisers mapped the territory, Angolan soldiers shot Soviet weapons, Cuban troops protected Angolan soil, and Cuban doctors treated the war's wounded. The term "Cold War," or "Guerra Fria," never circulated innocently in the press. It was never the term of choice to describe the power interests at play in the region. Announcers might quote or ironize the term, but they never used the words directly. Instead, they employed the term "imperialism" to describe the ongoing political struggles in the region.

Global and alternative histories of the Cold War have reimagined the era from the archives and experiences of those countries and peoples whose histories do not fit an East-West story of conflict. Many underscore how imperial projects condition these new "worlds" (Third World or "Darker Nations" in Vijay Prashad's formulation, for example).[13] Christopher J. Lee poses decolonization during the Cold War as an entrance to a political scenario continuous with the colonial era, not a clean exit from those relations.[14] In southern Africa, white supremacist politics and the Cold War hindered and conditioned decolonization of the region's countries. Lee calls this "decolonization of a special type" to mark continuities with apartheid that leftist thinkers from the 1950s on described as "colonialism of a special type."[15] The ALCORA Exercise, a secret military and political pact set in 1970 between apartheid South Africa, Rhodesia, and Portugal to maintain white sovereignty in southern Africa amid and against independence in

FIGURE 5.1 RNA's logo (past and current). The image contains the machete and cog from the Angolan flag that announce the country's socialist program and, today, that history. The tower radiates that commitment, and RNA's call sign in the 1980s announced the station as "the firm trench of the revolution in Africa." Used with permission of RNA-E.P., Luanda, Angola.

the other countries on the African continent, extended what Maria Paula Meneses and Bruno Sena Martins call "colonial dreams." When the Cold War heated up in the southern Africa region, "various imperial projects were also in confrontation."[16] ALCORA evidences ongoing colonial intentions in these fragments of empire, white-dominated holdouts on an otherwise sovereign continent. Colonial and imperial logics persisted. They collided head-on with nationalist revolutionary regimes in Angola and Mozambique. The effects of ALCORA in shaping what Vladimir Shubin calls the "hot Cold War" and on politics played out in postindependence civil wars, migrations, and regional political games.[17]

Recent scholarly work demonstrates the complexities of superpower and Cuban actions in the southern Africa region.[18] Onslow asserts that "despite the preoccupation of politicians and the security forces in Salisbury and

Pretoria of an all-embracing Soviet assault on the sub-continent, in reality, Soviet and Cuban involvement only accelerated in response to South African, and American backed, pre-emptive action in Angola in 1975."[19] Disagreements ruffled secret and not-so-secret alliances.

Flores's song and Mbete's introduction of Castro underscore that South Africa and Angola's "entrances" into sovereignty and independence thundered with the support of new alliances—the Afro-Asian Alliance, the Tricontinental Conference, and the Organization of Solidarity with the People of Asia, Africa, and Latin America (OSPAAAL)—that also carried the marks of reimagined imperial projects as much as those of new Third World collaborations. National liberation movements and their supporters (like the Organization of African Unity's Committee for African Liberation) experienced as much conflict as cooperation.[20] Nationalist movements sought ideas and material resources. Fighting to free themselves from colonialism, movement leaders did not want to submit to new, external forms of intellectual or economic control. They found Marxism and socialism appealing because these ideologies and politics offered solutions to the extreme economic inequities that characterized colonial economies and politics and a way to understand them.[21] Just as liberation movements believed that Marxism and nationalism could quickly transform the unequal trading and political relations between Western economies and underdeveloped ones, white settler states believed that socialist revolutionaries would remove them from their positions of political and economic power. White isolation, a product of white supremacist thinking, produced a siege mentality, while nationalist revolutionaries, often infiltrated by counterinsurgency informers, saw conspiracies to prevent their liberation at every turn.[22] In both cases, local nationalisms found external resources in Cold War and Third World ideologies to bolster their programs.

Whether Marxism, nonalignment, or anticommunism, national liberation movements, their resultant political parties, and rebel movements reached outward for ideas and alliances that they thought would help them realize their political ideals and ambitions. Clapperton Mahvunga conceptualizes southerners as political engineers who weaponize themselves with northern tools for postcolonial projects of their own design.[23] This is as true for weapons used on the battlefield as for those employed on the airwaves. In this case the weapons are both technological (radio) and intellectual (propaganda).

Between 1962, when Algeria secured its independence, and 1975, when the MPLA declared an independent Angola, the polarizing dynamics of the Cold War narrowed political options. Nonalignment was not viable in southern Africa. The Third World alliances imagined and forged at meetings like Bandung or in the Pan-African and Pan-Asian alliances of the mid-twentieth century persisted. MPLA leaders and party militants still revered Yugoslavia's Marshal Tito.[24] MPLA guerrillas, like their Algerian counterparts, staged media events and played off regional and global rivalries to assert their sovereignty, win diplomatic recognition, and impose their hegemony at independence in 1975 (with the help of Cuban troops). This was how they entered and managed the geopolitical world through decolonization. We might imagine these actions as part of a larger set of strategies that decolonizing and postcolonial countries developed in the second half of the twentieth century. The MPLA called this out directly, emphasizing the continuity of imperialism in Cold War politics and discourse.

Here I draw your attention to Angola's insistence—what today may be too easily dismissed as strident, socialist agitprop—that the ongoing South African colonial occupation of Namibia impeded full Angolan independence and that South Africa's military incursions and economic manipulation constituted imperialist aggression against the Frontline States. It was not merely agitprop. This was a reasoned argument that deployed the terms of international law and grew out of the experience of invasion by multiple forces. This argument aimed to shape a narrative about Angolan independence as part of a regional project. Its repetition on the radio and, years later, as a memory of the good Angola did in the region, evidences the power of radio to shape a story.

MASS MEDIA AND POSTINDEPENDENCE PROPAGANDA

The state endowed RNA with the largest budget of all media and let radio professionals purchase the best material available on the international market instead of demanding the least expensive pricing or preferential contracts with Eastern Bloc suppliers.[25] In Mahvunga's terms, they "pulled in" the best technologies to support the radio and build a national network that extended far beyond what the colonial state built.[26] Fighting a civil war and subject to foreign invasion, the state needed a vast and nimble communications network to blanket the territory with its sound and broadcast beyond its borders. The MPLA party/state required the technical capacity to make the political argument to its population and to those listening in from afar while it fought the war on the ground. As had been the case for the late colonial and metropolitan states during the war, this was a question of the right line (propaganda) and the right technology (radio) to send out the message. In the civil war, the experience of invasions fundamentally shaped these priorities. The pressure on transport and the heightened state of defense meant efficient nonroad forms of territorial communication, such as the radio, took on new signficance.

Of all the media in Angola, RNA had more resources, better facilities, and bigger staff than *Jornal de Angola* (the state daily), Angop (the newswire), or TPA (the television station). This had to do with the historic development of radio in the country as much as the privileged mobility of sound over text in a country where orality, and aurality, predominate. Prior to the first elections in 1992, press freedom arrived first for print media and then slowly opened to other nontext forms. In the years just after independence, radio was the most significant form of media. Paulo de Carvalho and Reginaldo Silva explain:

> During the First Republic, Angolan National Radio held a monopoly on radio in Angola, broadcasting in shortwave, medium wave, and FM to the whole national territory. It was (and continues to be today) the only communications medium with broad national coverage, which can also be heard in neighboring territories. RNA was used by the MPLA (the political party that runs the Angolan state) to deliver its official discourse and political and economic messages to the country's citizens. It was a difficult period for journalists at the radio due to party control over the radio's message.[27]

The regular presence of RNA and radios in Angolan literature of the postindependence generation underscores this.[28] Though the MPLA confidently announced that for the five-year period ending in 1985, "Rádio Nacional de Angola confirmed its place and role in the National Information System, establishing itself as the mass medium with the greatest reach and penetration in the country," no audience statistics exist for this period.[29] The state's monopoly on radio broadcasting obviated the need to measure listnership.[30]

Radio was part of a multipronged mass media strategy. MPLA party policy stated that mass media—radio, cinema, television, literature, and the press—"complete and generalize the work in party structures in information and orientation."[31] In other words, mass media would spread information and engage in political, cultural, scientific, and aesthetic education of the "masses" in the period of National Reconstruction that defined political life after independence. Party policy put it this way: the press "cannot confine itself to the task of relating isolated or chronologically ordered facts but must offer an interpretation of the causes and consequences so as to contribute to the political and ideological education of the masses."[32] Educating listeners, framing the story for them, and forwarding a narrative were the media's charge.

MPLA PROPAGANDA

Paulo Jorge, Angola's foreign minister from 1976 to 1984 and the man responsible for setting up *Angola Combatente*'s second station in Lusaka, said, "In Algeria I learned to kill people with a pencil."[33] This was not an admission of macabre murder techniques; rather, Jorge revealed something about how propaganda worked in national liberation struggles on the African continent during the Cold War. To amplify the impact of a guerrilla action, he recalled, "one would report casualties in an odd number and in quantities slightly above those actually recorded."[34] The MPLA had sent Jorge to Algeria where, in addition to training new recruits, he learned the art and artifice of radio propaganda from Algeria's FLN (Front de Libération Nationale). Matthew Connelly argues that Algeria's FLN innovated a "foreign policy of national liberation" that used media, international organizations, the politics of nonalignment, regional tensions, and emerging Cold War dynamics to win official international recognition of FLN leadership and support for Algerian independence even though they failed to establish territorial control.[35] The FLN provided a model for liberation struggles in decolonizing Asia and Africa. As Jorge's story shows, Algeria's FLN actively supported national liberation movements like the MPLA. Jorge reported the tactics of news propaganda, but the MPLA likely learned about strategy as well.

The MPLA charted a new course for the nation, tapping into the practices of decolonizing countries and socialist countries. Delinda Collier observes this dynamic in the Angolan cultural policy of the "Homem Novo": "The formalized New Man represents the branding of a methodology: the scientific socialist method of state formation circulated within an increasingly networked set of countries."[36] At the same time, those who produced arts policy, communications, or foreign policy adapted communist theories to local realities.[37] This happened inside RNA, as the last chapter argued. In state propaganda, the party/state interpreted general communist principles about the role of mass media and ideas about international capital relative to local conditions and regional circumstances.

The MPLA constructed its position vis-à-vis Namibian independence in the international press, in radio news flashes, and in reports to the UN aimed to mobilize the Angolan population and to galvanize international opinion. This strategy developed over time. Propaganda after independence had to look and sound different from propaganda for the anticolonial war. Indeed, MPLA documents recount a history of mass media that defines the period of anticolonial war as distinct. MPLA party documents from the First Party Congress in 1977 discuss the history, role, and significance of MDM (mass media).[38] The party/state quickly nationalized the press, radio, and television station (the TV first broadcast after independence).[39] It prioritized building infrastructure for transport and communications: "In a country the size of Angola, the existence of an efficient communications network that allows for the rapid establishment of contact between different regions and between Angola and the world is indispensable."[40]

Policy documents highlight the critical role of national and international news in "political education," spreading the principles that guided Angolan foreign policy. Mass media should underscore the work of the national liberation movements on the African continent that fight all forms of racial discrimination and should highlight Angolan solidarity with those movements.[41] News reports should focus on the social and economic victories in socialist countries, on the struggles of workers and peasants in and outside Angola, and on how imperialism, neocolonialism, and fascism oppress populations.[42] These documents used international socialist vocabularies that circulated around the globe in the 1960s and 1970s. They also expressed local priorities and desires: independence, regional liberation from white settler states and their white supremacist politics, and participation in that struggle.

While Paulo Jorge received training in propaganda during the guerrilla struggle, others landed positions in information and propaganda posts at moments when necessity thrust those with communications skills and education into the public eye and in front of a microphone, as was the case for "Comandante Jújú" (Júlio Almeida). Others, like Paulino Pinto João ("PPJ"), the director of the MPLA Department of Information and Propaganda (DIP) from 1983 to 1989, had no communications training. That innocence made him useful to the party leadership in the late 1980s. President dos Santos appointed PPJ in the wake of a debacle known as "*a peça do quadro*" (the painting play), in which a number of key intelligentsia lost their posts in the party, including in the DIP. PPJ suddenly found himself at the pinnacle of the DIP.[43] Unprepared for the role, he proposed to the MPLA that he complete a six-month course in propaganda and journalism in Cuba before assuming the reins of the DIP in 1983. MPLA propaganda that had developed from experience in guerrilla struggle, from bases outside the country, was not necessarily the best model for propaganda once the party held the state. One particular case proved critical in the transformation.

Propaganda responded to contingency. It was policy in motion. Though it ranged across media, radio sounded out a significant key. The MPLA leadership promoted those in charge of propaganda and radio in the guerrilla phase (Adolfo Maria, Mbeto Traça, Paulo Jorge) to other posts after independence.[44] Between the military coup in April 1974 and the declaration of Angolan independence on 11 November 1975 by the MPLA in Luanda, the three different movements fought for control of Luanda. CIA money covertly supported both the FNLA and UNITA, South African troops entered the country on a secret mission, and the MPLA relied on the support of Cuban troops and Soviet tanks and weapons. Once the MPLA declared independence, the battle had only just begun both inside the country and for international recognition. The ruling MPLA used media, namely radio and to a lesser extent television, to broadcast its claims of sovereignty to audiences inside and outside Angola.

In a savvy move to have its rule legitimized by foreign governments, the MPLA state played on the capture of US and British mercenaries and regular South African troops to impugn the involvement of the apartheid government and its Western allies in the newly sovereign territory. One month into independence, in December 1975, the MPLA captured thirteen mercenaries. The state tried them in a highly publicized trial in 1976. Keen to garner worldwide attention, the MPLA organized a legal team in the Ministry of Justice to try the case. That team courted the international press, inviting them to Luanda and live-translating trial proceedings in four languages. RNA broadcast the whole trial live.[45] *Time* magazine covered the Luanda Trial of 1976 in a short piece titled "Death for 'War Dogs'" in their 12 July 1976 issue. Journalist Wilfred Burchett and law professor Derek Roebuck used the trial as the launchpad for their saucily titled diatribe, *The Whores of War: Mercenaries Today*, written in close consultation with the MPLA, and published by Penguin Books in 1977.[46]

Angola's People's Revolutionary Tribunal sentenced four of the thirteen to death by firing squad: Daniel Gearhart, age thirty-four (a US citizen recruited via an advertisement in *Soldier of Fortune* magazine); and three Englishmen: the notorious Costa Georgiou, age twenty-five (a.k.a. Colonel Tony Callan, accused of killing thirteen of his own men and ordering countless tortures and deaths of innocent Angolans); Andrew McKenzie, also twenty-five (second in command); and John Derek Barker, age thirty-five. The other nine received sixteen- to thirty-year sentences in Luanda's São Paulo prison.[47] The tribunal declared "death to international imperialism and its lackeys!"[48] Aside from the international publications, a contemporaneous Angolan publication, *O Povo Acusa/Julgamento dos Mercenarios/a Legalidade Revolucionaria* (The People Accuse/Trial of the Mercenaries/Revolutionary Law), published in Luanda in 1976 describes the trial.

At the invitation of the Angolan Ministry of Justice, a forty-two-member International Commission of Inquiry on Mercenaries (composed of lawyers, judges, ministers of justice, academics, and journalists) witnessed the proceedings and declared them fair. During the trial, the People's prosecutor Manuel Rui Monteiro, a celebrated Angolan writer, derided the US as "the home of the CIA and the mother of mercenaries" and called Henry Kissinger "the international crime syndicate's traveling salesman."[49] The prosecution targeted the US and British governments through the mercenaries. Speaking at the trial, and recorded in *O Povo Acusa*, Monteiro noted that the only defendant without a right to a defense lawyer was imperialism itself.[50] Secretary of State Kissinger and the UK's Queen Elizabeth appealed for clemency for the accused. Neto refused; the executions went forward. According to Burchett and Roebuck, this case had significant international repercussions—the mercenaries embodied the growing danger of secret wars and the trial showed the need for public scrutiny and debate.[51] A solidarity group in England, a relic of the anticolonial struggle, worked to advertise the trial proceedings, unearth information about the London-based mercenary recruiter, and support the work of Burchett and Roebuck.[52]

Around the same time the Angolan military captured the mercenaries, in mid-December 1975, they also caught four South African soldiers in Kwanza Sul Province. Part of Operation Savannah, a clandestine South African military endeavor to prevent the MPLA from seizing Luanda and declaring independence, these soldiers donned Portuguese military camouflage and were instructed to speak with English accents in order to pass themselves off as mercenaries.[53] Captured two hundred miles from Luanda, their location belied South African claims that they entered Angola only to secure the Ruacana Dam in Cunene Province, some one thousand kilometers to the south. Holding the line, South African radio broadcast then defense minister P. W. Botha's claim that these four men were logistical technicians sent to fix a broken-down truck who had lost their way and strayed into MPLA territory.[54] The MPLA took full advantage of the opportunity. Commandant Júlio Almeida "Jújú" interrogated two soldiers on RNA on 17 December before exposing them to the international press. Evidence of this did not come from the basement of RNA but from the military archives of South Africa, where military brass listened in. The SABC (South African Broadcasting Corporation) did not, however, broadcast this news. At least not immediately and not precisely as the military had received it. Reception was good, retransmission less so.

Radio was the first but not the most effective medium for this story. South African historian Gary Baines notes that in South Africa, "it was the visual imagery that did the most damage to the apartheid state. A photograph of the four dejected young white South Africans was given worldwide exposure. Their body language suggests shame, even disgrace. The Angolans scored another propaganda coup when two of the four captured soldiers were displayed at a press briefing in Lagos, Nigeria, on 18 December 1975."[55] The image of these SADF (South African Defense Force) soldiers—white, handcuffed, in the hands of sovereign, black African authorities, inverted the racist power hierarchies that had made South Africa a pariah state. Although Fred Bridgland, then the Reuters correspondent, had reported the presence of regular South African troops in Angola as early as November, the international press held silent.[56] South Africa now had to come clean about its operations in independent Angola. The last South African troops withdrew from Angola in March 1976. The Soweto student uprising followed in June. Angola and Mozambican independence in 1975 and the capture of South African troops paraded before the international press exposed South Africa's secret war and inspired antiapartheid forces.[57]

After the capture of South African soldiers and US and British mercenaries and the MPLA's successful handling of them in the media and in diplomatic circles, the OAU (Organization of African Unity) recognized the MPLA as the rightful and independent government of Angola in February 1976. The captures and the propaganda around them raised the relationship between the visual and the aural. While the aural—for all its intimate, ephemeral, hard to fix and locate characteristics—disturbed the PIDE, it did not provide the kind of sticky, visual evidence that captured soldiers did. Interviewing them on radio proved less powerful than seeing them in the hands of newly independent, black African leaders (at least for what it meant in South Africa), though likely at home too.

Radio played a part in spinning the capture of the mercenaries and South African soldiers into evidence of imperial invasion, but it was the visual evidence that was damning. In graphic, visual terms, it demonstrated MPLA authority. The MPLA used the captures as an opportunity to articulate a narrative about outside forces threatening Angolan sovereignty and meddling in Angolan politics. The captures were time-stamped events, something to hang a chronology on. Radio—through broadcasts that created continuity and presence—maintained that narrative line throughout the late 1970s and 1980s. The consistent framing of these issues in news flashes and presidential speeches, like the one from Huambo, created a state ideological frequency that pushed back against control by other global geopolitical discourses of the Cold War, capitalism, and apartheid's claim that Angolan independence under the MPLA represented the global creep of communism.[58]

The mercenary trial and the capture of the South African soldiers offered highly mediated events in which the MPLA developed its propaganda strategy. The party crafted propaganda by drawing on the vocabularies of international solidarity, international law, and revolutionary rhetoric to trace imperial continuity across the supposedly clean break of independence in the actions of Portugal's erstwhile allies and current competitors for access to Angola's markets and resources (namely the United States and South Africa). The propaganda charted change inside the territory, now governed by the MPLA, a liberation movement with a socialist revolutionary project. While the visual imagery of the mercenaries exposed South Africa and explicitly inverted regional power dynamics, radio broadcasts brought the sounds of war—the rush and buzz of airplanes—and the language of international diplomacy and solidarity into Angolan homes and workplaces. It kept the narrative about foreign intervention and imperial machinations present in the listeners' ears.

RNA NEWS FLASHES

News flashes formed a key part of daily radio broadcasting. This passage from Ondjaki's *Bom Dia Camaradas* (Good Morning Comrades) highlights the radio's central place in the home and some of the ways it shaped thinking:

> We were kind of bored with the news because it was always the same thing: first, news of the war, which was almost always the same, unless there had been some important battle or UNITA had blown up some pylons. That always got a laugh because everyone at the table was saying that the UNITA leader, Savimbi, was the "Robin Hood of the pylons." Afterwards there was always a government minister, or someone from the political bureau, who said a few more things. Then came the intermission and the publicity for the FAPLAs. Oh yeah, that's right, sometimes they talked about the situation in South Africa, where the African National Congress was. Anyway, these were names that you started to pick up over the years. Also, you could learn a lot because, for example, on the subject of the ANC, my father explained to us who Comrade Nelson Mandela was and I found that there was a country named South Africa where

> black people had to go to their houses when a bell rang at six o'clock in the evening, that they couldn't ride the bus with other people who weren't also black, and I was amazed when my father told me that this Comrade Nelson Mandela had been a prisoner for I don't know how many years. That was how I came to understand that the South Africans were our enemies, and that the fact that we were fighting against the South Africans meant that we were fighting against "some" South Africans because for sure those black people who had a special bus just for them weren't our enemies. Then, also, I saw that, in a country, the government's one thing and the people are another.[59]

The radio's consistent sound and ideological constancy might produce boredom, but the radio kept the war present, and it interpreted the Angolan civil war in regional, geopolitical terms. In doing so, radio created "affective rhythms" that structured daily life and offered a reliable narrative throughline in the quotidian chaos produced by war.[60] Remediated, as in this passage, and in conversations with friends and family, news gained affective force.

The use of regional geopolitical terms was not new. Such language predated Angola's independence and intensified after it because it helped to name the politics of white supremacy at work in the region. South Africa and Portugal had allied to protect imperial and white settler interests in the face of the liberation movements that turned to armed, anticolonial struggle against Portugal in Angola and Mozambique. Their alliance started as early as 1963 with mechanisms of cooperation in military intelligence, but it sounded deeper notes from 1966, when the MPLA opened the Eastern Front (contiguous with southern provinces where SWAPO could operate from 1965). The South African military began to lend a small amount of military and logistical support during that period.[61] The 1970 ALCORA Exercise joined Portugal, South Africa, and Rhodesia in a tight, white, secret embrace and deepened the relationship. After independence, South Africa suppurated the wounds in the region by supporting UNITA in Angola and RENAMO in Mozambique, initially formed by Ken Flower and ex-PIDE agent Oscar Cardoso.[62] White supremacy was not afraid of black allies if it meant holding on to power.[63]

Even if ALCORA had not yet been exposed, it was not difficult to see that South Africa's continued occupation of Namibia threatened Angolan and Namibian sovereignty. RNA news flashes thus hit consistently on the necessity of Namibia's independence. They decried South Africa's illegal occupation of Namibia, determined by the International Court of Justice decree of

June 1971 and endorsed by the UN Security Council in October of the same year. The OAU reiterated that call. World opinion concurred. RNA broadcast this until it was common sense. Taking recourse to international governing bodies and international law, MPLA propaganda gave local events, such as bombings of southern Angolan cities, world significance.

Angola's independence, in the eyes of South Africa's apartheid regime, constituted a threat to its security. Both SWAPO and the ANC, MPLA allies, would set up bases in southern Angola.[64] In Piero Gleijeses's summation, from the South African state's point of view, "the MPLA had to be destroyed."[65] UNITA leader Jonas Savimbi, on the other hand, was keen to collaborate with the apartheid regime, and the CIA and South Africa willingly supported him with matériél, logistical support, and troops.[66] After the exposure of Operation Savannah in 1975, the Angolan army, supported by Cuban troops, routed the SADF on 27 March 1976. But this was a Pyrrhic, if symbolically rich, victory. Even as the UN condemned South African aggression and demanded compensation for damages done by the invasion, new attacks commenced in late June 1976.[67] Indeed, the UN Security Council, between 1976 and 1979, passed resolutions that condemned South African attacks on Angola five times. Repeatedly, the US, France, and the UK abstained.[68] Just as repeatedly, RNA broadcast these votes in news flashes.

In 1978 the UN Security Council passed Resolution 435, reaffirming United Nations legal responsibility over Namibia, calling for a withdrawal of the South African administration, a cease-fire, and transfer of power to the Namibian people via UN Transition Assistance Group–supervised elections.[69] That would happen only eleven years later in 1989. In the meantime, the South African war against SWAPO and Angola continued. The UN mission tasked with collecting information on South African acts of aggression committed after 11 June 1979 concluded that attacks hit civilian targets in Angola. Sparing neither women, children, nor the elderly, these bombings constituted "crimes against humanity" and an "undeclared war against an independent country" for repeatedly violating territorial integrity and national sovereignty.[70]

The Angolan authorities sent a report to the UN. It argued in the following terms: "We believe that, with the unveiling of these new crimes, we will be able to awaken international public opinion, and especially the French people—who were the brain of world revolution—to the fact that all the South African crimes against our nation have been perpetrated with war material made by South Africa under French licence."[71] They appealed to French national political pride (and its inverse: shame) to compel not only French condemnation of South African aggression but to call out France publicly on the international stage for not just aiding and abetting South Africa but for profiting from apartheid too.

Inside Angola, RNA reported on this diplomatic exchange regularly. News flashes have not been systematically archived, but for the early 1980s they are consistent. South African invasions and attacks featured regularly in both local and international news flashes. Crafting Angola's place on the continent and in the world, RNA endeavored to inform, educate, and mobilize Angolans. In RNA broadcasts, Angola and Angolans emerged in defense of their territory and in the anti-imperialist and antiapartheid fight. The RNA news flashes knit Angola into the region and the world as an independent nation-state whose defense of its sovereignty had regional and global effects.

The news flashes drew a straight line from the MPLA-PT's First Party Congress's "Theses and Resolutions," which defined the work of radio in propaganda and in educating the Angolan people. The party underscored the importance of mass media to national reconstruction and the construction of socialism. Following Marxist-Leninist doctrine, mass media should serve national development and education.[72] More concretely, the MPLA-PT said that radio, television, and the press should "give particular support to the militant and effective solidarity of our people with the fellow peoples of Africa and especially the southern African national liberation movements."[73]

To this end, RNA reported on international and UN support for the antiapartheid movement and how apartheid had an effect on Angola. The visit of the UN mission, headed by Paulette Peirson, headlined the 11 p.m. local news flash broadcast on 6 August 1980:

> The chief of the International Commission of Inquiry into the Crimes of the Apartheid Regime in Southern Africa, Paulette Peirson, was in Luanda. She condemns the South African invasions in our country. The General Confederation of Work in France expresses its will to contribute to the end of acts of aggression by Pretoria against the People's Republic of Angola.[74]

With only the pause of punctuation to separate them, these items flow seamlessly into one another. This news flash produces a sonic cordon of protection generated from UN pronouncements and French worker solidarity. Messages like the following brought worldly embraces to Angola's situation and amplified a sense of international concern: "Today there were demonstrations of Vietnam's solidarity against South Africa's aggressions."[75]

When invasions ramped up in 1981, another flurry of broadcasts highlighted the territorial breaches, official condemnations, and protests against them:

> The Minister of Defense denounces South African terrorist actions against the free and sovereign territory of the People's Republic of Angola. Various reconnaissance flights by South African aviation have flown since April over Xangongo, N'Giva, Cuamoto, Mavengue and Mamcunde and between Lubango-Moçamedes. Bombings have caused 5 deaths and 27 wounded, both civilian and military; 2 deaths and 25 of the wounded, all civilian, were caused when they bombed two buses on the Xangongo-N'Giva road. UNTA and the Confederation of Free Syndicates in Democratic Republic of Germany call for a solution to the Namibian question via resolution 435 and support for SWAPO—that will be communicated today on the program "Work and Struggle." João Hailonga, the OAU rep for the Committee of Freedom in Angola, condemned the South African racists. In a meeting on the 18th anniversary of the foundation of the OAU he spoke of "the necessity for Western countries to abandon a politics of defending their warlike interests and hypocritical behavior based on the total disinterest in the decolonization of the African continent."[76]

Key points often reiterated throughout the news cycle included civilian targets, Namibia and UN Resolution 435, the OAU, decolonization, and South African racists and their Western allies. Breathless reporting of this sort transmuted the local to the global, the micro to the macro. UNTA, the National Union of Angolan Workers, was tied to the East German Confederation of Workers, grammatically and sonically realizing their international brotherhood, abolishing distance and difference with a conjunction. The flash never mentioned UNITA. RNA left UNITA's program and politics in silence, devaluing the party's platform by ignoring it. When news flashes focused on international players, they offered the MPLA-PT position that South Africa and its Western allies were the real enemies. UNITA was a lackey not worth mentioning.

Radio news broadcasts not only framed events but sought to mobilize Angolans. Local news exhorted the population to turn out for protests against the invasions. Reporting from Lubango, the site of constant reconnaissance missions and occasional South African bombing, RNA covered a protest organized for the following day in late August 1981: "Adelino Catimba, member of the MPLA-Workers' Party Provincial Committee, who will preside over the act, encouraged all the men, women, and youth to be prepared to exchange their tools for arms in defense of our free and sovereign territory."[77] Ten days later a protest in Luanda took place:

> The population of Luanda will protest against the South African invasion. The MPLA-Workers' Party provincial committee of Luanda exhorts workers and all the people of the province to show up at 9h00 in front of the Ministry of Defense to march to the May 1st Plaza. The secretary of National Union of Angolan Workers exhorts all workers to participate. A holiday has been conceded between 8h30–12h30, work piquetes will operate where necessary.[78]

These exhortations were featured between news of musical performances, economic sabotage, adjustments in currency, and the paralysis of bread production.[79] They helped build the sounds of socialism and nationalism in local life.

The international news emphasized regional questions with geopolitical repercussions: the Frontline States, debates at the UN, solidarity with the ANC and SWAPO, Nujoma's international travels, and relations of Western states with South Africa. In mid-April 1980, the UN Security Council analyzed Zambia's official complaint about South African attacks. Angola's ambassador at the UN, Elísio de Figueiredo, speaking in the name of the Frontline States, "underlined that the minority regime of South Africa is in a panic over the role played by the Namibian people, led by SWAPO, in undertaking its liberation, by the call for struggle by the South African people, and by the recent victory of the Zimbabwean people."[80] Drawing together struggles in the region, he occluded the deep tensions between Kaunda and the Angolan state (over the latter's funding of UNITA) and instead pointed to the nervous, trigger-happy condition of minority rule in South Africa.

RNA covered SWAPO president Sam Nujoma's speech before the Frontline States in Maputo following a surge in South African destabilization in 1982, noting that Nujoma "reiterated SWAPO's disposition to negotiate a cease-fire and the application of the UN Security Council's resolution 435 that foresees free UN-supervised elections in Namibia."[81] These radio news items kept listeners' ears attuned to imperial aggression, blurring the stark polarities of the Cold War with the old motives of empire in the new guise of Western powers using covert operations to support a racist state. They took recourse to international law to defend not only the right of

Namibia to independence, but Angola's right to defend itself and to call on the assistance of Cuban troops to do so: "The Political Bureau of the MPLA-Worker's Party blames Pretoria for the state of War and rejects the connection made between the removal of Cuban troops and the independence of Namibia."[82] In this way, RNA news flashes interfered with the frequency of Cold War propaganda and discourse by refusing its terms. This was a strategy the MPLA honed in the trial of the mercenaries: anchoring an argument about emergent state sovereignty in international institutions and international law.

Another set of radio news flashes broadcast indirect and direct forms of counterpropaganda. News from the Nigerian press figured regularly in this category, providing a way to attack US imperialism on the continent. An announcer read: "A Nigerian newspaper denounced the presence of 70 North American information centers that coordinate the ideological expansion of imperialism on the continent. The VOA defames young African states. The CIA has agents in information organizations. The formation of PANA will help to eliminate the information monopoly of the USA in Africa."[83] Another news flash announced that the *Nigerian Herald* decried the West's neocolonial strategy in which cultural imperialism was a key tactic and Nigeria a prime target: "They are attempting to intoxicate Nigeria with the psychosis of the 'Cold War' against the Soviet Union and for that, they take recourse to all means, from the press, to radio, to cinema, the paper pointed out."[84] Here, the *Nigerian Herald* refers to the Cold War in scare quotes and describes it as a psychosis, a loss of contact with reality, a delusional state in which violence may occur. And on 3 August 1980, announcers slipped in a little note from the US press: "The *NY Times* analyzes the situations of blacks in the US which declines every year because of racial discrimination, socio-economic inequalities, and the total indifference of the authorities."[85] Here US claims to promote democracy abroad are undone by racial injustices at home.

Responding to the increase in South African aggression in 1982, RNA news flashes contradicted reports on South African radio:

> South African radio said yesterday that there was an operation of Cuban troops in central and southern Angola. The Angolan ambassador Mawete João Baptista in Havana contradicted this and recalled the bombing of Cahama in Cunene by the South Africans the week before. He called for a peaceful solution to the Namibian question and "as to the establishment of diplomatic relations with the USA" he said it would only be possible as long as the principles

of non-interference in the internal affairs and the sovereignty of each state were respected.[86]

International news on Namibia and South Africa and the crisis of decolonization came intercalated with pieces about the MPLA government and politics in other countries. Mid-June 1980 found the then director of Propaganda and Information, Comrade Manuel Pedro Pacavira, traveling to Cuba to exchange experiences with his homologue regarding "ideological work and mass media."[87] In November 1980, RNA announced the election of Ronald Reagan, "an ex-actor of the cinema. In the 1940s Ronald Reagan acted against actors of cinema in the McCarthy witch-hunts. He promises to increase the military budget to negotiate with the USSR. He refuses a minimum wage for workers."[88] Announcing Reagan's nonpolitical background, hawklike position on the Cold War, and capitalist commitments in succession served to ironize these elements and underscore the threat he posed.

The fighting in southern Angola continued. The 1988 battle of Cuito Cuanavale proved decisive to political negotiations. Negotiations that led to the New York Accords in December 1988 had in fact already begun, but the military situation advanced them. Scholars, military generals, and diplomats have weighed in on the significance of this "battle."[89] They chart the conflicts over operations between Cuban, Soviet, and Angolan military advisers, on the one hand, and South African and UNITA military advisers, on the other.[90] Considered the largest land battle since World War I on the continent, in the end, it was not a battle but a stalemate in the face of tremendous military mobilization.[91] Gleijeses described it as a "defensive victory" based on Cuban air superiority and Cuban and FAPLA successfully repelling a South African attack east of Cuito.[92] The outcome was not military bloodshed (though Angola's civil war certainly cost many lives), but South African military withdrawal. The strong and destructive South African artillery had inflicted damage in Cuito Cuanavale and air raids had proven lethal in southern Angolan cities. But South Africa would withdraw from Namibia and the USSR and Cuba would withdraw from Angola. Namibia moved to independence through free, UN-supervised elections. If some South African generals have been keen to contest the idea that South Africa lost, one scholar framed it this way: "In terms of the propaganda war, it need hardly be said that South Africa was the loser, particularly since the diplomatic necessity of a blanket news silence covering the entire episode."[93]

In the aftermath, the MPLA-PT burnished Angola's military and political credentials. Reflecting on Namibian elections and their effects on peace in

Angola, President José Eduardo dos Santos, interviewed by CNN (through a Cuban translator), said that the main determinant of regional stability would be the elimination of apartheid. Asked about who the prime movers were in catalyzing the agreement on Namibia, dos Santos pointed to a combination of agents and a change in international forces. But Angola played the determining role (Cuba was a secondary agent, in his reading) by directly and indirectly resisting South African aggressions. Angola supported the struggle of Namibian people for independence. Angolans suffered losses in human lives and economic losses in the range of $20 billion, the president noted. With due encomiums for the US and the USSR, he underlined Angola's various methodologies and initiatives for achieving peace on the question of Namibia over the years.

The CNN interviewer then asked: "Do Gorbachev's policies of enlightened socialism (glasnost and perestroika) have any effect on your thinking in Angola?" President dos Santos responded: "Before the USSR initiated its restructuring process, Angola had already announced its economic reforms. Our second congress which defined the new economic policy for Angola was held in December 1988, and it seems that the 22nd congress of the Soviet Communist Party happened later."[94] President dos Santos and the MPLA segued from criticisms of the West to calculated negotiations, Angolan agency always foregrounded. The West (and the East) might have imperialist ambitions; they might start wars, invade, create trouble. But Angola did not follow, it led. This narrative strategy kept Angola at the center of the story, despite its reliance on Cuban troops, material, and advice, as well as on Soviet arms and counsel. In this telling, Cubans and Soviets were supporting actors in an Angolan drama. Change in the Soviet Union might in fact follow, or at least come after, change in Angola.

Opening to the market in the late 1980s—in fact before 1989—and not in line with superpower Cold War chronologies and the end of the civil war in 2002, turned state interests to questions other than the war with South Africa, especially once Namibia gained independence.[95] While Namibia lingers in people's memories, the Angolan state does little to retain it as a strand in its own history. Since the end of the war in 2002, the official narrative has become future-oriented and pivoted to national reconstruction and growth. The once glorious alliances of the past are the subject of demure anniversary celebrations and the occasional monument. State media and politicians accuse those who raise the issue of the civil war (in whatever guise: reconciliation, Namibian independence, protection of sovereign territory from South Africa, the rights of veterans, etc.) of interfering with a long-awaited peace. This is interference with the new frequency of state power and the new narrative of postwar Angola. State power is ever more centralized in the executive and little interested in the socialist promises of yore.

The radio figured as one part in the Angolan state's propaganda endeavor, particularly in broadcasting to its own population, and in supporting SWAPO and the ANC with their radio programs. Earlier chapters charted radio as a constant of the Angola soundscape (since the 1930s), so how listeners received these frequencies varied, though it is largely outside the purview of what I can recover. I did not encounter reports on range or reception at RNA, though the creation of the Department of Information and Public Opinion is a reminder that radio personnel thought about audiences.[96] Songs like "Semba pra Luanda," literary works like those by Ondjaki and other writers, and photographic, visual, and performance works by South African and Angolan artists show the impress of the war and the mark of propaganda on framing a narrative about Namibia and Angola in the 1980s and 1990s.[97]

6 Radio Vorgan

A Rival Voice from Jamba, 1979–98

TOWARD THE end of Angolan writer and journalist Sousa Jamba's novel *Patriots* (1990), UNITA's Vorgan radio station enters the narrative. The character Osvaldo refers to "Kwacha radio," a reference to Vorgan. Osvaldo is an MPLA soldier; his brother Hosi is a UNITA soldier. A house divided, a family fighting itself, is a microcosm of a country riven by ideological difference and political interest. Referring to Vorgan, Osvaldo says: "Pity, Radio Kwacha seems to have broken loose some people's nerves."[1] He is suggesting that Hosi's uncritical listening to Vorgan has led him to paint a hostile picture of MPLA-ruled Angola. A fictional representation, this passage is noteworthy for returning the language of nervousness to the discussion of radio listening and its effects after independence. Vorgan made listeners nervous, and not just by creating a paranoia among UNITA's members. It made the Angolan state nervous and it made international listeners, namely those involved with the peace process in Angola, nervous too.

This chapter listens for the history of Vorgan over its nearly twenty-year existence, from the station's beginnings in January 1979 until its last broadcast in April 1998.[2] Despite the fact of political differences that spurred and fed a twenty-seven-year civil war with the ruling MPLA, a history of UNITA communications—itself a history of radio and aspirations to state power—sounds similar to that of the MPLA. Radio played an important role in UNITA's propaganda and in the lives of UNITA members and militants. UNITA spun true lies. It made some listeners nervous and it assured others.

Jonas Savimbi, founder and longtime president of UNITA, played international and local audiences to great effect. Like the MPLA at the height of the Cold War in the region, he lost no opportunity for media spectacle. Grand claims about military incursions, imminent victory, and truculent threats composed the daily fare of UNITA's press. Vorgan, the "Voice of the Resistance of the Black Cockerel," UNITA's clandestine station, was a cornerstone of that media outfit. By the late 1990s, the UN accused it of continued virulent attacks on the MPLA-led government and of acting in bad faith relative to the Lusaka Peace Protocol. Using one party's propaganda to damn the other's (drawing on MPLA descriptions of Vorgan), scholar Elaine Windrich referred to Vorgan as "a laboratory of hate."[3] But what more can be said about it? Interviews with former Vorgan journalists, monographs on the central highlands and UNITA, and anecdotal accounts of listeners portray a more complicated story.

Vorgan broadcast a new voice for the movement that Jonas Savimbi founded in 1966. Vorgan was part of a guerrilla war effort against the MPLA government. Like the party it sprang from, the station was illegal from the point of view of the MPLA-controlled state, making mobility and secrecy paramount. The Federal Broadcast Information Service (FBIS), noting the source of information, always described Vorgan (and Kwacha UNITA Press, the UNITA newswire) communications as "clandestine." Eventually, Vorgan set down roots in the bush in Angola's southeast, as part of UNITA headquarters in Jamba. But the significance of information and communications to UNITA lies not only in war. As recent research on Angola's central highlands and political history demonstrate, a more tangled and nuanced set of relations pertains between education, modernization, technology, and the community of this region, and so too is the case between those threads and UNITA as a guerrilla movement.

A SKETCH OF RADIO AND THE *PLANALTO CENTRAL*

The literature on nationalism has pointed repeatedly to the influence or relationship between mission education and the development of nationalist thought and movements, whether the FNLA, MPLA, or UNITA, in Angola.[4] Recent social histories shed light on how missions shaped quotidian life, producing new forms of education, leisure, and worship.[5] Diverse cultural practices thus characterized colonial and mission-shaped life on the *planalto* (plateau) by the mid-twentieth century.

Protestant missionaries, and in particular the Congregationalists based at the Dondi mission station in Huambo, worked to realize a godly kingdom in a rural world, safe from the depredations of urban industrial life that plagued the West and the assaults of Portuguese colonial society delivered in the name of modernity.[6] Missionaries imagined modern African villages and productive agriculture manifesting a divine kingdom. Their charges, like those analyzed in a vast recent literature, took up this call and interpreted, reread, and rewrote it in a variety of different ways.[7] The Congregationalist/Protestant communities of central Angola opened rural and urban bases by the early to mid-twentieth century. By the 1950s, Protestant missionaries and church elders fought to keep mission-educated young people from migrating to urban centers and joining the civil service and colonial milieu (one example of the ways Africans did as they saw fit with missionary education and cultural modernizing projects).[8] Mission education (whether Protestant or Catholic) provided upward social mobility and an entrance to the colonial world.[9]

Conceição Neto points out that "despite the aims of the missionary organizations, the strategies of their Angolan flocks often pointed to other directions and both Catholic and Protestant Umbundu elites tended to identify 'urbanization' with 'modernization' and a 'modern' advanced education with better living standards and social status."[10] Catholic missions, and the Spiritan seminary that trained a "native" elite, believed that "spiritual threats" permeated urban life.[11] Whatever temptation and sin awaited in towns, they attacked it directly. They opened the Kanye Mission near the city of Huambo in the 1930s, and in 1947, they founded a seminary in Huambo's city center.[12] Because only seminaries provided access to the coveted "civilized"/ *assimilado* status, even Protestants converted to Catholicism sometimes.[13] Whether or not one joined the priesthood, a seminary education placed one among the intellectual elite of the colony, even if segregation and racial discrimination continued to set limits on employment.[14]

Didier Péclard uses Jean Bayart and Stephen Ellis's idea of "extraversion" to read Ovimbundu engagement with mission schools as part of a longer historical trajectory.[15] Peoples of the *planalto central* (central plateau) from the era of the slave trade up through the present have looked and linked themselves outward, with other peoples in Angola and across the Atlantic. In caravans across the interior, through the years of rubber trade, in their relations with missionaries and the colonial state, and in the period of the liberation struggles and beyond, plateau kingdoms and societies sought connection.[16] The invention, adoption, and use of communications strategies and information technologies is part of this long-term practice. Like the personnel at RNA in the last chapters, societies of the planalto and UNITA as a political movement are "usage engineers," reaching out and drawing in techniques and technologies from afar (from East and West, on and off the continent) to serve their own ends.[17]

As accoutrements of the colonial world, radios marked civilized status on the plateau as they did elsewhere in colonial Angola. Machines that connected one to the rest of the colony and the rest of the world, to others in a literate Christian ecumene spanning the globe, radios signified modernity in their very technology because they materialized the invisible flows of power, learning, and networks that made broadcasting possible. They were part of what Neto described as the dominant white culture of the settlers and missionaries: "new names, dress styles, house furniture, eating habits, etc., all working as signs of distinction."[18] The habitus or worldliness of whiteness that Ahmed describes shaped interactions on missions and in towns. This was the colonial cultural context from which urban life on the plateau and UNITA grew.

Sousa Jamba's family was part of this cultural world. His novel *Patriots* mentions radio more frequently than any other work of Angolan fiction. Published in English by Viking in 1990, the novel's semiautobiographical narrator Hosi is an educated, cosmopolitan exile who returns to Angola to fight for UNITA in the civil war. Jamba's story ripples with the energy of various media (kung-fu films, radios, turntables, and music).[19] Radio broadcasts (BBC, *Kudibangela*, Radio France International, RNA, SABC/RSA, Transworld Radio, VOA, and Vorgan—called Kwacha Radio) and radios appear as purveyors of national, and international, news and propaganda.[20] Jamba, trained as a journalist, has a keen ear and eye for the work of media in everyday life. Radios sing out news, often distorted by political ideology, but the persistent desire to know, to take in information, interpret, and rethink it, is in constant tension with the official forces that want to enforce conformity.

VORGAN: VOICE OF THE RESISTANCE OF THE BLACK COCKEREL

A shrewd military strategist, a keen manipulator of people, a double-talking charismatic leader, Savimbi micromanaged every detail of battle and life in his movement.[21] Information and communications were no exception (see fig. 6.1). Savimbi entwined informal and formal systems of information collection and distribution. Within the movement, Savimbi employed gossip, for example, to manage relations among the leadership and to destabilize the most intimate relations between rank-and-file members. He used spies to rat out dissenters.[22] He sought intimate information to manipulate those

FIGURE 6.1 Jonas Savimbi, leader of UNITA, 10 August 1978, Angola. Photo courtesy of Keystone Pictures USA/Alamy stock photo. Used with permission of Alamy.

closest to him.[23] Savimbi invented a story about Samuel Chiwale, once a top UNITA leader, that led to Chiwale's social disgrace and demotion in the UNITA hierarchy.[24]

UNITA's leader used information to consolidate and maintain power, to control the movement leadership, and to project UNITA's claims to the outside world. Savimbi used information in military, political, public, and private affairs with an understanding of its capacity to unsettle as much as to shore up individuals and institutions. He invested in formal information systems. Radio figured in UNITA propaganda strategy. Along with Kwacha UNITA Press and the newspaper *Kwacha*, Vorgan broadcast UNITA's presence and philosophy outside its military deployments and its headquarters in Jamba. Savimbi saw radio and information as a crucial part of the war with the MPLA. Former Vorgan reporter Lourenço Bento described the military ends of communications: "Combat was not just fought with arms but with radios," at times, literally. Savimbi went so far as to distribute radios so that villagers could listen to Vorgan.[25] And the state sometimes confiscated them, as Kwacha UNITA Press reported in November 1988.[26]

Windrich, and peace process observers in later years, focused on the political vitriol and grandstanding in Vorgan broadcasts. Vorgan targeted government lies and bad will. Vorgan denounced international diplomats, once seen as loyal allies, who made what Savimbi considered errors of judgment in negotiating the transfer of power in Namibia or peace in Angola.[27] Broadcasts trained their sonic crosshairs on movement members and leaders in bad standing.

Savimba set up UNITA headquarters in Jamba in the southeast of Cuando Cubango Province along the border with Namibia in late 1979.[28] Vorgan began broadcasting in January 1979, its earliest transmission captured by the British Broadcasting Corporation before the movement settled in Jamba later that year. The station first broadcast, not announcing its location, from a base in South African–occupied Namibia as "the Voice of the Good Black Man's Resistance."[29] Journalist Bela Malaquias remembered a system of recording at the Delta Camp in Namibia's Caprivi strip. They recorded programs there and then sent reels to Windhoek for evening broadcasts.[30] But Windrich notes that neither the BBC *Summary of World Broadcasts* nor the Foreign Broadcast Information Service recorded Vorgan broadcast activity again until 1983 when a more regular schedule, or at least more regular transcription in the West, commenced.[31] At that point, they had chosen the name of the "Voice of the Resistance of the Black Cockerel" (having also broadcast as the "Voice of Truth" and the "Voice of the Resistance of the Black Rebellion").[32]

One Vorgan announcer recalled that Vorgan first broadcast from cars—one a white Mercedes sedan, the other a brown Unimog—outfitted with small-scale broadcasting equipment provided by South Africa during Vorgan's early days in Jamba.[33] Fit for war, they could move precious broadcasting equipment quickly and report while on the offensive or in retreat.[34] Eventually Vorgan acquired 10 kW transmitters located at more permanent studios and could broadcast directly.[35] The United States provided container studios that South African technicians set up.[36] By the mid-1980s, the movement had constructed a building dedicated to information services. It housed Vorgan editorial headquarters, the offices of Kwacha UNITA Press, and the party's graphics office, though broadcasting continued from the containers.[37]

New transmitters increased the station's strength tenfold. UNITA did not have a way to systematically study or follow listenership.[38] Anecdotal information they happened upon gave them some sense of their audience and broadcast range. For example, UNITA broadcasters learned that Vorgan could sometimes be heard as far north on the continent as Morocco (a UNITA ally). This report returned with Jorge Valentim, who had participated in the negotiations preparing the Lusaka Peace Protocol of 1994. At those meetings he received documents showing that Vorgan had been heard in North Africa and by the Voice of America listening center outside Monrovia, Liberia.

Reports about listening from inside Angola were more common. UNITA guerrillas and populations across the country tuned in. Bento heard stories of clandestine listening, not unlike those recounted by people who listened to *Angola Combatente*—hiding in rooms, under covers.[39] Broadcasting in shortwave with a 10 kW transmitter and a system of antennas, the signal could be picked up widely both within and outside Angola.[40] Bela Malaquias said that even today people recognize her voice from having listened to Vorgan during the war years. Her mother, in exile in Canada in those days, knew Malaquias was alive because friends in other parts of Angola sent word that they had heard her reading the news on Vorgan.[41] Miraldina Jamba broadcast under the name "Kouela Jamba." Returning to Luanda after 2002, she met a nurse who worked at Américo Boavida Hospital who named his daughter after her radio pseudonym. He told Jamba that he liked the name and had followed the show regularly, if secretly, in Luanda.[42] In Huambo, Jamba's mother also listened clandestinely. In a 1988 piece titled "UNITA Radio Causes 'Great Concern' for MPLA," Kwacha UNITA Press reported that the broadcasts reached as far north as Cabinda and into neighboring countries. Likewise, they claimed, the MPLA had been known to arrest listeners and interfere with radio and battery purchases, though they could not verify "reliable sources" (a repetition of the colonial state's nervous behavior relative to the guerrilla radios).[43]

The MPLA-ruled government listened intently. RNA recorded Vorgan and Vorgan and RNA engaged in a war of the radio waves. This was the first rule of propaganda: engage the enemy's information and invert it. Angolans in exile in Zambia tuned in and, as Miraldina Jamba's story of the girl named after her radio pseudonym demonstrates, some Luandans pricked up their ears as well. People tuned in, even if they did not agree with UNITA, simply to hear a different version of events than that reported on RNA.[44] Despite the desires of broadcasters, listeners were not just echo chambers; they recognized propaganda, "heard between lines"—discerning more or differently than intended—and they could play an RNA broadcast against a UNITA one.

One Vorgan journalist estimated that at its height, Vorgan had a network of five hundred journalists working throughout the country in the areas it controlled (with Jamba as the station's headquarters): editors, reporters, operators, and technicians.[45] Correspondents accompanied troops to battle and the directorate on political and diplomatic missions. This staff received training both within and outside Angola. Savimbi presided over and directed the station, but it fell almost immediately within the remit of the information secretary and a station director oversaw Vorgan's day-to-day operations.[46]

Current UNITA president Isaias Samakuva held the post of radio logistical director in Jamba for a period. The directors ran the local training courses and supplemented them with workshops with visiting journalists and technicians such as South African Ricardo Branco.[47]

RICARDO BRANCO AND THE SOUTH AFRICAN CONNECTION

Ricardo Branco is a man shrouded in mystery. Mentioned by name by at least three former Vorgan journalists as a trainer seconded by the South African military through Radio South Africa (the external service of the South African Broadcasting Corporation, SABC), sometimes for periods as long as a month, Branco denies being anything but the translator for another South African journalist.[48] His story illuminates some of the complex dynamics of the postcolonial Portuguese world in the southern African region. Perhaps because he spoke Portuguese, Vorgan journalists could communicate with him directly and remembered him specifically. Branco's skills and experience as a journalist—working throughout the region, breaking the news of Samora Machel's death to the Portuguese press, training journalists at the SABC, covering the Inkomati Accords between Mozambique and South Africa, as well as establishing and running Channel Africa for the SABC's Radio South Africa—would have made him a memorable figure.[49] Those attributes also would have made him the perfect person to do the training (not just the translating). He may have been the interpreter. He may have been the trainer. Or both, at different times.

Branco got his start in radio at the well-regarded Rádio Clube de Moçambique (RCM). Like the radio clubs in Angola, it was amateur and membership-based. Unlike the clubs in Angola, it was the only one in Mozambique. Large and well-organized, RCM broadcast in the territory and the region. While many Portuguese returned (or went for the first time) to Portugal to escape the turbulent decolonization, others relocated, especially to South Africa (or to other white settler colonies like Australia or the US).[50] South Africa had a sizable population of "white/not quite" Angolans and Mozambicans who exercised a white skin privilege they were largely accustomed to in the former Portuguese colonies, though within a heightened segregationist climate in apartheid South Africa. Given the regional political situation, such people could take up important political and military roles. Former PIDE agent Oscar Cardoso is another such figure.[51] Though it is not clear what connection there might have been with the disbanded ALCORA Exercise, it is clear that the political and cultural ties and sounds of whiteness persisted in the region. Someone like Branco, with broadcasting and language skills, would have been indispensable for communicating South Africa's propaganda in southern Africa. Like Malaquias, who did not love politics but loved broadcasting, when asked about politics, Branco too takes refuge in professionalism. In a region charged with the racial dynamics of decolonization, the struggle against apartheid, and crisscrossed with Cold War interference, doing your job and doing it well offered security. Being true to one's professional and technical skill, under such extreme circumstances, might mean broadcasting "lies."

VORGAN PROGRAMMING

Professionalism offered psychological space to Vorgan journalists. Vorgan programs, as journalists remember them, broadcast material beyond what Windrich referred to as "hate," and outside the political timbre that caught the attention of FBIS monitors and transcription specialists. This points to the cant of the archive (in this case, the FBIS) and to how those who worked at the station found meaning in everyday work, with or without political conviction. FBIS transcriptions do not mention music, by and large, but Vorgan broadcast a wide variety, from Roberto Carlos, to Whitney Houston, to Zairean music, to Antillean music.[52] UNITA offices overseas purchased this music and sent it to Jamba to blare cosmopolitan sounds even from the bush.[53]

Malaquias had a morning show, *Good Morning Africa! Good Morning Angola!*, to rouse and inspire, and a nighttime show geared toward relaxation. Educational shows ran during the day: motivating students with study skills while pedagogical courses offered content. Each area of the movement had its own program: LIMA (the women's organization); JURA (the youth group); Alvorada (for children); and SINTRAL (Free Angola Workers' Syndicate—for workers). Vorgan broadcast the main news in Portuguese but also translated it into eleven African languages and into French and English.[54] Journalists developed all aspects of the show themselves, including writing and music. This allowed them to "lie in the abstract with true things in the particular."[55] Shows that meant something to them, in which they took an authentic interest and worked hard, might support UNITA's "lies." That was the price of survival.

Lourenço Bento composed part of a four-person team that put out the daily show *Pátria Livre* (Free Country), UNITA's Forces for the Armed Liberation of Angola (FALA) program. *Pátria Livre* presented UNITA's political doctrine, explained the political purpose of combat, and aimed to motivate soldiers. Sound technicians at the studio recorded music from UNITA bands

Resistência Negra (Black Resistance); Negros Oprimidos (Opressed Blacks); and O Som da Liberdade (Sound of Liberty), and this music featured frequently on the show.[56] In one segment of every show, soldiers exchanged messages—greetings from soldiers in Cabinda to those in Cunene, or a salutation from one soldier in Benguela to another in Moxico. This was an important and particularly popular aspect of the program that boosted morale.[57] *Pátria Livre* also engaged in the propaganda game with the state military. The show's announcers hailed soldiers drafted into the FAPLA or forcibly recruited in *rusgas* (roundups at night or at schools) and urged them to desert.[58]

Broadcasting evolved with political and military strategy. Raúl Danda introduced a program in Ibinda/Fiote (one of the main languages of Cabinda) at Vorgan in 1985, after UNITA invited him and four other Cabindans to Jamba and then forcibly prohibited them from leaving. A move to open a front in Cabinda, all four received military training and schooling at the Centro de Estudos Comandante Kapusa Kafundando. The UNITA directorate then sent three of them to Cabinda to establish UNITA there. Danda remained and marched through Cuando Cubango, Cunene, Huíla, Moxico, and Bie, as was the practice for UNITA soldiers. On return to Cuando Cubango, Franco Marcolino invited him to Vorgan to present the program in Fiote for broadcasts to Cabinda.[59] The program eventually became part of the national languages service. Danda, for his part, transitioned into doing programs in Portuguese and live broadcasts. He ascended through the ranks: director of services in Ibinda; editorial director; head of international services; and finally to the position of director, under Jorge Valentim as information minister and Norberto de Castro as vice minister.

Journalists based their accounts on direct reporting and reading the information collected by Kwacha UNITA Press (KUP).[60] KUP produced information on political policy and military actions. Journalists had access to news from international news agencies: BBC, VOA, Deutsche Welle, Radio France International, and newswires like Reuters, Lusa, and Agence France-Presse. Other sources included UNITA's armed forces (FALA), including information they had intercepted from FAPLA and government messages they decoded.[61] The media area had a satellite dish and television. This allowed them to follow television news from on and off the continent.[62]

Journalists had a small margin in which to maneuver between the restrictions of self-censorship and the party's directives. Danda proposed that correspondents report from different areas of the country, even if they were not there. Another kind of "true lie." The news would come from soldiers via military communications radios and the announcers would set up in the studio but create a different, on-the-site sound.[63] Confirming this practice, in January 1994 RNA issued a direct attack calling out false reports from correspondents said to be in other provinces. For the state, these were nothing but "lies":

> The so-called correspondent would be sent into a room, where he was issued with a poorly tuned recorder, and then he had to shout into it a text cooked up in a hurry by people such as [UNITA Information Secretary] Jorge Valentim and company, at the end of which he would say: this was such and such, reporting from Huambo. First, that such and such was not in Huambo and, second, that is not his right name. . . . We saw poor Amos, who is also known as Bob, and whose duty it is to listen to other radio stations, doing the job of correspondent in the room next to where he works. We saw the wretched Carlos, a white man who uses the pseudonym of (Catjabala), trying to pass as the station's correspondent in Bie.[64]

This motivated Savimbi to establish live broadcasts from different areas.

Miraldina Jamba ran the LIMA radio program. Trained in journalism in 1982 at Jamba, her background is in education. Unlike the other journalists I interviewed, all of whom were younger when the decolonization process began and the civil war broke out, Jamba joined UNITA in Europe as a student, newly married and persecuted by the PIDE. She and her husband, Almerindo Jaka Jamba, fled Portugal for Switzerland to present themselves to the UNITA representative there (Ruben Sanjovo Chitacombi) in 1973.[65] In 1974, when the revolution in Portugal on the 25 of April inaugurated the decolonization process in Angola, Savimbi asked UNITA youth outside the country to return. The Jambas went to Luena, where Jaka Jamba was elected UNITA delegate. From there Miraldina Jamba broadcast the program *Kwacha* at the Rádio Clube de Moxico from 11 a.m. to 12 p.m. They wrote and read news, communications or announcements on the war, and played music.[66] The program moved to Luanda during the government of transition when EOA conceded broadcast time to each of the three political movements. Jamba and Ribeiro Chitaculo worked as the announcers. Their tenure there was brief.

In August 1975, Norberto de Castro and Maria Dina, already at the EOA, fled with Jamba and others from Luanda to Huambo when the relative calm of the transition fragmented into violence and a struggle for territorial control.[67] They set up broadcasting at the Rádio Clube de Huambo when Savimbi and Roberto declared the Democratic Republic of Angola there. News at 1 p.m. and 8 p.m. and entertainment programs constituted the main material, supplemented by political programming. Jamba experienced this as on-the-job training with seasoned journalists Norberto de Castro and Maria Dina, who had worked for years at EOA.

In February 1976 UNITA fled Huambo for Bié, then moved to Cuando Cubango, then to Moçamedes, and eventually into the bush for years, before settling permanently in Jamba in the early 1980s. Once they opened the Vorgan studio, Jamba created the program *Página da Mulher* (Women's Page) in her capacity as LIMA information secretary (from 1984 to 1991). The show aired three times a week for an hour. The segment format included an editorial statement (political in nature), interviews (with various women), and educational pieces (on subjects like family planning, health and hygiene, education). In this sense, LIMA's program differed little from that of OMA, the women's organization of the MPLA, and its modernist, developmentalist vision of women.

If broadcasting from the EOA as part of the transitional government and then from the Rádio Clube de Huambo during the four months in which UNITA controlled the cities of the central highlands made UNITA sound like an heir to the colonial state, a rightful inheritor of sonic sovereignty, broadcasting as Vorgan and from Jamba was quite another endeavor. Vorgan exemplified many of the characteristics associated with the kind of state UNITA hoped to embody. It may be for precisely this reason that some former Vorgan journalists seamlessly transitioned into work at Rádio Nacional de Angola at various points after 1992.[68] Vorgan was a professional operation. Radio journalists were UNITA cadres, the well-educated movement foil to the peasant base whom they were meant to represent and who were the recipients of the movement-supplied services. Cadres peopled Jamba's institutions. The peasantry provided soldiers, food for troops, and weapons transport for UNITA. UNITA leaders used this distinction between cadres and peasantry to justify the capture of professionals taken to Jamba.[69] Some came voluntarily, others by force or under conditions somewhere between the two.[70] As to the radio journalists, Miraldina Jamba joined the movement outside the country and returned voluntarily; Raúl Danda (and his colleagues from Cabinda) was coerced and then stayed, Bela Malaquias was swept up in the movement out of the central highlands cities in February 1976, Lourenço Bento joined at the age of fifteen, and Armando Ferramenta attended school at Jamba from a young age.

A high degree of political and social control permeated life at Jamba and other UNITA camps.[71] Journalists faced constraints: "A person would engage in self-censorship even for physical safety. We'd give flight to our imaginations but within that defined space."[72] "I'm not a political person," Malaquias said.[73] Malaquias focused on learning her craft and perfecting her skills. Bento, Jamba, Danda, and Ferramenta, without the same disdain for politics (indeed, Jamba and Danda at the time of our speaking were both

UNITA representatives in Parliament and Bento the UNITA president's secretary), also emphasized the professionalism of the work and the capacity to learn and grow as journalists within the ranks of Vorgan. Similar to RNA, the politics of control made professional focus a potent survival strategy.

From the collection of the FBIS (starting in 1983), it is clear that much of what Vorgan broadcast (or much of what FBIS transcribed) fell into the category of war reportage—casualty counts, battle reports, ideological framing (e.g., "Russo-cuban invasion"), interviews with MPLA "deserters," ripostes to official RNA news (and the other official state mouthpieces like *Jornal de Angola*, Televisão Popular de Angola, and the ANGOP newswire), and more elaborate military and political situation reports.[74]

Sometimes one broadcast covered many bases. A 1983 interview with two so-called deserters (it would be hard to know if they deserted or were forcibly recruited) hit on key elements of UNITA philosophy. On Vorgan they said, "Yesterday we were enslaved by the Portuguese, today we are enslaved by the Cubans," accused the MPLA leadership of robbing the people (calling them "vampires"), discussed MPLA raids on youth for the army, referred to Vorgan as the source of truth about FAPLA defeats, and encouraged their former comrades to desert.[75] Visits of foreign dignitaries and journalists received on-air note. Savimbi's speeches constituted another Vorgan staple.[76] In one of Savimbi's recorded messages, titled an "Address to the Nation," he said that in addition to the broadcast via military channels, "we will also ask Vorgan to continuously broadcast it to the entire country," blanketing all of the territory with Savimbi's notion of nation.[77]

One broadcast, from March 1985, took on declarations made by the MPLA on the tenth anniversary of Angolan independence. This broadcast underscored elements of UNITA political philosophy and propaganda, occluding key past events that told a different story. The broadcast noted the presence earlier that month in Jamba of "dozens of journalists who represented international public opinion."[78] Pre-dating the repeal of the Clark Amendment but subsequent to US president Ronald Reagan's public proclamations of support for UNITA and Savimbi, this press gambit aimed to demonstrate UNITA's historic role and political coherence before an international audience. The Vorgan broadcast, however, opened with an address to the local audience: "Compatriots, comrades, Angolan people."[79] Again hitting the note of "Soviet-Cuban imperialist and neocolonialist interests," the broadcast referred to the MPLA ruling government as "the puppet and blood-thirsty regime of Luanda" and as "chieftans who are presently massacring Angolan people with the assistance of the invading Cuban troops . . . to maintain Angolan blacks as slaves."[80] Here the broadcast implied the MPLA's illegitimacy based on its creole urban roots, its lack of cultural authenticity from the point of view of UNITA. It was not a party of black Angolans, and it was a party complicit with colonialism. This opened space to make a claim for UNITA's historic role in the anticolonial fight: "It is a well-known fact that without UNITA there would be no victorious struggle of the Angolan people against Portuguese colonialism."[81] In fact, it was Savimbi who in the name of UNITA had signed a secret pact of military collaboration with the Portuguese military and against the MPLA in 1971.[82]

Finally, the broadcast struck notes that would resonate in the West even as they played up UNITA's international status for a local audience. The MPLA's rule constituted "an embarrassment to the free nations of the world" and had set Angola back despite the fact that it possessed the infrastructural and technical conditions "for a solid take-off on the path to progress."[83] Nothing like Walt Rostow's own phrase to assure then current (South Africa, France) and potential future (US) funders of your capitalist spirit. Such broadcasts demonstrate that Vorgan hailed multiple audiences—local, foreign, state, and popular—and addressed them with different discourses and ideologies, often in the same broadcast.

By the late 1980s and early 1990s, numerous broadcasts detailed the ins and outs of peace negotiations in Gbadolite, Zaire; Bicesse, Portugal; and Lusaka, Zambia. Reading translated transcripts of these broadcasts highlights radio's technopolitical aspects. For example, a team of German parliamentarians visited Jamba in December 1990. The broadcast noted: "The delegation has come to the capital of freedom to learn about the outcome of the meeting between Dr. Savimbi and Soviet Foreign Affairs Minister Eduard Shevardnadze, and recent developments in the peace process, as well as UNITA's program for an Angola in peace. Yesterday afternoon, the team toured the Information Secretariat, during which Mr. (Schroeder) was interviewed by Vorgan."[84] UNITA used the radio installations to demonstrate their technological capacity to foreign visitors. Indeed, not just the radio station but the Information Secretariat. Radio constituted a part of techno-institutional space meant to showcase UNITA's fitness to rule. This movement had national ambitions and the institutional capacity to back it up.[85] At the same time, if the interview had been broadcast (though I could not find the record of it), it would showcase the movement's diplomatic work and Savimbi's international stature to Angolan and foreign audiences. While earlier reports often brought delegations to witness military actions, like the 1983 attack on Munhungo township in Bié Province that was "filmed by a team from an international delegation," the emphasis over time shifted to curating technical and bureaucratic expertise for international viewers.[86]

Radio mattered not just for broadcasting outside but as part of daily life in Jamba. People used radio intensely in the bush.[87] All party officials had radios.[88] People listened collectively during the main news broadcast hour. Radio carried messages of encouragement to soldiers on the front and it provided contact with the world outside.[89] Isolated in southern Angola for strategic purposes, radio provided a thin line of access to friends and family at other bases and in exile. At headquarters and at camps, Jamba and Bento remembered that UNITA leadership encouraged militants to listen to a variety of stations.[90] Malaquias recalled a more restrictive set of listening practices. Group listening posed a risk. Savimbi, ever keen to control and use one person or group against another, had informants everywhere.[91]

THE STATE OF JAMBA

The third movement to emerge among the nationalist liberation movements, UNITA remained the smallest in political and military terms until after independence. When the civil war broke out, it garnered the external support of South Africa and mobilized parts of the population of the central highlands, energizing political commitment, drawing on circumstance, and using force to become a viable movement.[92]

The UNITA/FNLA union in November 1975 dissolved by late December in recriminations and battles in the southern cities of Sá de Bandeira and Moçamedes that continued through January.[93] The January 1976 debacle with British mercenaries, the CIA, and Zaire, discussed in chapter 5, sealed the FNLA's downfall and neutralized them politically.[94]

Under siege by state FAPLA and Cuban forces in February 1976, UNITA set off on what party members and others refer to as the "Long March."[95] In a calculated move, UNITA abandoned the cities of the central highlands and took to the bush when the South African military withdrew support and SADF forces beat a steady retreat in February and March 1976. As the Brazilians had in November, Western diplomats recognized the People's Republic of Angola upon the advancement of government troops.[96] UNITA found itself isolated and unable to hold any cities. The movement desperately needed someplace remote to implant their political, military, and social institutions. Government and Cuban troops occupied the central highlands, great in population density, making UNITA's first choice for a base impossible. Two leaders scouted a location in Cuando Cubango that first became a reeducation camp for prisoners from the Delta Camp but soon developed into the movement headquarters.[97]

Even though in the late 1970s and early 1980s UNITA could make only circumscribed territorial claims, it had nationalist pretensions.[98] Jamba constituted the headquarters of the movement, a makeshift capital and place that projected "UNITA's aspirations to statehood."[99] Jardo Muekalia, who had left Jamba early on to represent UNITA internationally, stated, "As the capital of resistance, it represented the symbol of our force and the expression of our organizational capacity."[100] The men and women Pearce interviewed often remembered it in idealized terms—as a place where one could find food (local produce and imported goods), health care, and education. Miraldina Jamba described life at Jamba headquarters as a "small experience of normal life."[101] The location had schools, hospitals, the radio, zones designated for agriculture, an atelier for the production of uniforms, a mechanics workshop (for transport repairs, etc.), and a place to repair weaponry that doubled as a studio for specialists in ivory and wood sculpture.[102] These workshops served to recuperate weapons and vehicles captured in war and to produce crafts given to foreign visitors.[103] Photographs show a landing strip and large buildings, as well as smaller constructions where people lived.

If Jamba represented "normal life" for someone like Miraldina Jamba, who had grown up accustomed to urban ways and consumer culture, UNITA headquarters also operated as a war machine. Samuel Chiwale, former general commander of UNITA military, explained how it fit into Savimbi's "theory of big numbers" to create a nationalist mass movement.[104] Chiwale echoed Savimbi's logic: "The struggle against the Russo-cuban expansion will be a long one; it will take some time and many of these young people that are here will die or, in the best cases, be mutilated. Therefore, we have to find a region capable of regularly renovating our supply of guerrillas so we are never short of them."[105] Military capacity had political implications. Guerrilla units, a semiregular army, and finally a regular army meant that the movement had developed the basis for what Chiwale called "a State with a State" and an Angola with "two capitals."[106]

Despite its location in remote eastern Cuando Cubango, visitors and residents often referred to Jamba as a city. One of Pearce's interviewees associated UNITA's provision of goods and services with Jamba's urbanness. Fred Bridgland described arriving at "Savimbi's camp, a kind of spread-eagled forest city." A sign, written in English, announced: "Entering Free Angola."[107] Bridgland visited a secretarial school with twenty-odd students tapping away at typewriters under the guidance of a French teacher decked out in a Parisian suit and high heels; a hospital lit up by a Soviet generator where a doctor was performing surgery; and religious services at both Protestant and Catholic churches there.[108] A heavily fortified and densely secured set of giant huts, spread over hundreds of meters under the cover of trees, served as Savimbi's general headquarters. Bridgland remembers a number of radio operators poised before illuminated electronics equipment, allegedly confiscated Soviet supply, monitoring the radio and telegraphy communication flow among the MPLA's military.[109] On another visit he recalled sleeping "between crisply ironed sheets" and drinking "welcome cool beer in cans imprinted with the exhortation 'Keep South Africa Tidy.'"[110]

In a rare piece of criticism printed in the *National Review*, a mainstay of the conservative US press, Radek Sikorski, after a three-month visit with UNITA, described military battles, troop movements, and "bogus facilities" all staged for visitors and media. His excoriating tone continued: "soldiers and officials briefed on how to lie most plausibly—the full list of UNITA Potemkin villages would be too long to quote. Whether concerning the pettiest of details or the weightiest of policy matters, there is a great difference between UNITA's claims of high ideals, openness, and efficiency, and the reality of deceit, manipulation, and sloth."[111] Some forms of order did exist. Every fifteen minutes on the hour, Sikorski recounted, Vorgan came on with Mendelssohn's "Wedding March" and an announcement of Savimbi's greatness.[112] Pearce reads these displays differently. He sees a more intricate process of projecting authority both internally and externally: "It remains significant as an indication of the UNITA leadership's thinking: of how UNITA's leaders chose to demonstrate the movement's legitimacy in terms of its potential as a state and, moreover, as a particular kind of state"—modern, bureaucratic, service-providing, urban, and urbane.[113]

Calling Jamba the "capital of freedom" resonates with anticommunist Cold War discourses. One broadcast mentions that Savimbi met with Shevardnadze (this being late 1990), well after the fall of the Berlin wall and in the face of declining US support of UNITA. At this point, UNITA experienced pressure to negotiate with the government. No longer enjoying support from the US or South Africa (F. W. de Klerk's assumption of power spelled the end of regional intervention), after the negotiations around Namibian elections and South African and Cuban troop withdrawals at the New York Accords in December 1988, UNITA found itself isolated. It financed its war through artisanal diamond mining and an illegal international trade, in part set up by a former South African Defense Force officer who established contacts for UNITA in Tel Aviv and Antwerp.[114] The extension of UNITA's guerrilla operation to the northern areas in the mid-1980s witnessed initial incursions in the lucrative diamond mining regions of the Lundas, threatening government profits. After the Bicesse Accords, illegal, informal mining increased as diggers moved into the zone. UNITA, then in control of those areas, began to dominate much of that trade—profiting through taxation, selling licenses to diggers, and direct mining via foreign companies that used requisitioned labor—for the period between 1992 and 1998, before the government reestablished control.[115] Savimbi already traded in diamonds by the late 1980s, but the period after Bicesse was unprecedented in profits: "Industry sources put UNITA's cumulative net revenue in 1992–98 at around $2 billion . . . far more than the movement ever received from the United States and South Africa, its main external patrons, prior to the Bicesse Accords in 1991."[116]

UNITA bartered diamonds for arms and other military equipment or sold them for cash. UN sanctions against the purchase of UNITA diamonds, passed in 1993, proved difficult to enforce given the ease of smuggling diamonds. Savimbi created a web of connections between other African leaders (in Zaire, Togo, Cote d'Ivoire, and Burkina Faso) who offered end-user certificates in exchange for diamonds, international diamond traders, and a series of middlemen that made flouting the sanctions relatively easy and lucrative for everyone involved. Only as the war heated up again in 1998 did the Security Council enforce sanctions violations and enliven the Angola Sanctions Committee, which froze UNITA's assets.[117] All this technopolitical activity (i.e., mining for war economy and its strange bedfellows) was kept far from the microphones of Vorgan and from a public audience. That said, the MPLA-dominated state used this period after the election debacle and in the wake of a stalement around the Lusaka Protocol, a period popularly referred to as "nem paz nem guerra" (neither war nor peace), to tighten and extend its economic and political grip.[118]

STRIDENT SOUNDS AND INTERNAL DISCORD

If Vorgan's broadcasts often framed news in propaganda, for example, describing UNITA's participation in the 1989 talks in Gbadolite, Zaire, as the work of "our glorious movement," or closing with "Forward with the revolution," the tone of reports in the period around and after negotiations trilled with particular stridency.[119] Ad hominem attacks and vituperative commentary prevailed. African-led talks at Gbadolite produced a peace agreement that Savimbi denounced two months after signing it.[120] Fighting resumed. In May 1991 UNITA and the MPLA signed the Bicesse Accords in Portugal following negotiations over the course of 1990–91. It was during this period that Anita Coulson reported that the MPLA government referred to the station as a "laboratory of hate."[121]

The regional context had shifted. South Africa, focused fully on its own transition, no longer supported UNITA, and the USSR, post–fall of the Berlin Wall, had its own crisis to manage. The United States, for its part, preferred to negotiate with the MPLA for expanded access to oil rather than dither over political ideology. Vorgan broadcasts highlighted acts of bad faith in negotiations or countered state propaganda. For example, in February 1991 it responded to a Portuguese news (RTP; Rádio e Televisão de Portugal) item that reported on South African advisers in Jamba:

> In Jamba yesterday afternoon, Vorgan learned from sources in the UNITA foreign ministry that this news is ridiculous and groundless. Those sources said that José Eduardo dos Santos, the MPLA-PT's major ring leader, sent a letter to Frederik de Klerk, his South African counterpart, adding that UNITA sees it as an act of provocation that seriously endangers the Angolan peace process.[122]

Referring to the president of the Angolan Republic as a "ring leader," while not extreme, was a small verbal act meant to undermine the MPLA's construction of nation. Vorgan, and UNITA ideology and education, routinely reframed political leaders and their actions. Like Savimbi's "address to the nation," this constituted part of a larger ideological apparatus intended to communicate that UNITA conducted itself as a state, with authority, sovereignty, and legitimacy.

The negotiations at Bicesse broke down after the elections in September 1992. The MPLA won a majority of seats in the National Assembly (54 percent for the MPLA; 34 percent for UNITA; 12 percent for other parties); and in the presidential election, José Eduardo dos Santos won by a slimmer margin of 49.6 percent to Jonas Savimbi's 40.1 percent.[123] Without an absolute majority in the presidential race, a runoff would be necessary. That never came to pass. War resumed. Indeed, some of its most violent and destructive years occurred between 1992 and 1994 as UNITA seized cities and the MPLA capitalized on oil exports to reorganize its forces and rearm its military with new equipment.[124]

UNITA emerged with military successes after 1992, but it also suffered a number of stinging defeats over the course of the next many years. Rapid Intervention Police,[125] armed civilians, and government troops drove UNITA out of Luanda in November 1992, killing a number of key leaders and numerous UNITA supporters in attacks and raids in the city. Occupation of Huambo and Kuito backfired: "By the time UNITA retreated from the main cities in 1994, many of those who had quietly retained sympathy for the movement after 1976 had changed their opinion once they had experienced the way in which UNITA treated the towns and the people who lived in them in the early 1990s."[126] Important UNITA leaders defected. Miguel

Nzau Puna and Tony da Costa Fernandes simultaneously announced their departure from the party in 1992 in Lisbon. At the same time, they condemned the murders of Tito Chingunji and members of his family, making the news public for the first time.[127] Publicity about other human rights abuses, including the infamous witch-burning episodes that occurred in the early 1980s but came to light with Amnesty International's imprimatur in 1989, further discredited Savimbi's leadership and the movement more generally.[128] In 1994, the MPLA and UNITA signed another set of peace accords in Lusaka. Savimbi agreed reluctantly. He did not attend the signing ceremony himself and instead sent the UNITA secretary-general Eugénio Manuvakola.[129]

Defections from Vorgan started. Margarida "Guida" Sebastião Paulo left Vorgan and joined Rádio Nacional de Angola. In a press conference broadcast on TPA (Televisão Popular de Angola, Angolan People's Television) in mid-December 1992, she accused Savimbi of fomenting tensions between people of the north (like herself) and those from the south (predominantly Ovimbundu). She pointed to his abusive nature, discussed being physically beaten (saved from a worse fate only by the intervention of Nzau Puna, then UNITA secretary-general), and described incidences of Savimbi playing one leader off another. One case gained notoriety. Savimbi coveted the wife of a UNITA leader and punished the couple with long marches even during the wife's pregnancy.[130] At the same time, Paulo noted that UNITA needed to exist, that an opposition party helped make democracy work. She tidily communicated the MPLA position: negotiation with Savimbi was impossible, but UNITA was a viable, formal opposition party. This undermined Savimbi's authority but kept the party in play, even as it invited schism.

The couple Paulo mentioned, Eugénio Manuvakola and Bela Malaquias, suffered political and personal ordeals, as many in UNITA did, at the hands of Savimbi. This treachery paved the way for their defections. Savimbi sent Malaquias to Luanda in July 1992 (with her two youngest children) to cover the elections for Vorgan from a radio based in Largo Serpa Pinto. Claiming  that she had leaked information to the MPLA, Savimbi had her kidnapped and forcibly returned to Jamba, leaving her two children in Luanda. She feared for her life. Her brother, Assis Malaquias, began an international campaign to free her. This included an Amnesty International indexed call for letters to Luanda-based authorities in the Joint Commission.[131] In the wake of Chingunji's murder, international scrutiny focused on UNITA and Savimbi in particular. The elections brought renewed attention. Threatened repeatedly and finally dragged from her dwelling in Jamba in the middle of the night, Malaquias screamed and made a ruckus: "I didn't give them the pleasure of taking me quietly in the night."[132] UNITA showed her to the international press to demonstrate that she was still alive.

On election day, troops surrounded her home and prevented her from voting. "Savimbi was worried I would vote for the other party," Malaquias said. Eventually, UNITA transferred her to Huambo. When government forces took over the city she hoped to stay, but UNITA troops forcibly removed her and her family from their home (her husband was in Lusaka negotiating in Savimbi's name) and sent them to Andulo and then Bailundo. At the close of negotiations in Lusaka, Manuvakola returned with the Lusaka Protocol signed. He was greeted with jail time.[133] Savimbi denied authorizing Manuvakola's signature to the agreement. After another three years under UNITA control, they fled as a family in a group of fifteen people, from Bailundo to Huambo by car and then on to Luanda.[134]

Election propaganda worsened the timbre of broadcasting by both the MPLA and UNITA and, as journalist Ismael Mateus notes, "The resumption of war after the 1992 election triggered a resurgence of even more aggressive language."[135] Defections from Vorgan and Savimbi's refusal to stand behind his negotiators made the internal discord audible in public. In the meantime, Vorgan amplified its vitriolic attacks on the MPLA state and on those involved in the peace negotiations. State media participated too. Vorgan propaganda drew international attention for attacking UN peace negotiator Dame Margaret Anstee. A 1993 Vorgan broadcast accused Anstee of supporting the MPLA and refusing humanitarian assistance to people in Huambo. The broadcast used an ad hominem attack in grossly gendered terms, denounced the international community, and threatened her:

> The international community should be warned henceforth that in her adventures to supply FAPLA troops on the battle field, as was the case in Caimambo, on the day Anstee comes across a stray bullet—since a bullet does not choose—she will die like a FAPLA soldier and not like a UN official because of her greed. If the international community insists on having Anstee as a mediator in the Angolan conflict, then its own good will has been jeopardized, because what is at stake now is the nation and its people. In the first instance, this woman cannot have the caprice of leading the mediation process because she is corrupt. She is a prostitute—perhaps emancipated—who does not inspire confidence.[136]

The Joint Commission observers demanded an apology. RNA broadcast that demand in plain tones the following day.[137] UNITA broadcast an apology on the BBC from Abidjan, the site of the next round of negotiations mediated by Anstee between UNITA and the MPLA-dominated state. In that broadcast, the UNITA secretary for African affairs, João Marques Cacumba, apologized for the remarks and claimed that they had been made by a young man during the UNITA youth radio hour and not at the behest of UNITA leadership.[138] Disciplinary steps against the youth had been taken, Cacumba noted.[139]

SILENCING THE ROOSTER'S VOICE

Vorgan sometimes spoke in calm, statelike tones. Announcements like one by the Standing Committee of UNITA's Political Commission to propose peace "made an in-depth assessment" and suggested "direct negotiations."[140] Nonetheless, the bulk of Vorgan broadcasting rattled the nerves of those in government and some international negotiators. One source claims that the Angolan state went so far as to jam UNITA broadcasts.[141] Enlivening one of the ironies of the last stages of the war, in January 1996 UNITA announced on Vorgan that government-employed South African mercenaries were targeting the station.[142]

Vorgan broadcasts nettled the state. The agreements on radio written into the Lusaka Protocol set in motion the end of Vorgan broadcasting, the station's closure, its reinvention as a private FM station, and the opening of a UN mission radio station.[143] Lusaka required this under the rubric of national reconciliation, underscoring that Vorgan's bellicose sound had dangerous effects.[144]

Demanding Vorgan transition from shortwave to FM limited its broadcast range. A 1998 UNITA press release lamented that opening Rádio Despertar, "a non-party FM media outlet," would restrict UNITA's reception: "The Government has a monopoly on television and will soon have the only radio station, Rádio Nacional de Angola, to reach the entire country when Rádio VORGAN, the UNITA radio station, is replaced by the limited range of FM Rádio Despertar, a non-partisan media outlet."[145] RNA was the only national FM network.[146] Legislation that followed the Bicesse Accords liberalized the media around the elections. With many predicting UNITA would win, the MPLA feared losing control of state media. Figures closely associated with the MPLA opened all the new stations in 1991–92.[147] Vorgan stopped broadcasting in 1998. Rádio Despertar did not open until 2006. Media liberalization consolidated MPLA hegemony over the airwaves, instead of reducing it.

Epilogue

Jamming

"JAMMING" IS indirectly about radio. Here I briefly consider other media forms in the ongoing politics of communication in Angola. The state continues to control press, radio, and television, despite the liberalization of media laws in 1992. There is no such thing as dissident radio in Angola. Nonetheless, activists have challenged the state's monopoly on broadcasting ideas. Like the liberation movement militants before them, they take a page from political movements in North Africa. Angolans have turned to social media. This is not, however, a digital revolution. What is more noteworthy are the ways activists use new digital media with older, analog forms, like pamphlets, signs, and T-shirts, to "jam" the new state frequency of media power. They make noise.

For Jacques Attali, "noise is a resonance that interferes with the audition of a message in the process of emission."[1] These activists made noise when they disrupted the MPLA's postwar narrative that cast President dos Santos as the "architect of peace" and framed the MPLA as the best party to grow Angola's economy and better distribute its wealth.[2] Larkin calls noise an "unintelligible signal" that simultaneously excludes listeners and promises them future intelligibility.[3] Noise is cacophonous, nonharmonious sound that the ear cannot decipher . . . until it is not. That is the story of this epilogue: how jamming (interfering) with the state message became jamming (improvising music and creating a common narrative, noise's opposite). Noise turned music.

First, a discussion of what happened with radio between 1992 and 2002 before looking at the work of activists and more recent changes in contemporary Angola, including the introduction of new media laws in 2017.

MEDIA LIBERALIZATION AND RADIOCRACY

In the run-up to the 1992 elections, and as a piece of the transition to the Second Republic, the government passed the country's first ever press law.[4] In theory, this disentangled the media from the state and secured the right to inform and be informed. The MPLA, worried that it would lose its monopoly on state radio to UNITA should the opposition win, needed a plan.[5] In retrospect, it was just a step in what Didier Péclard calls the MPLA's "authoritarian reconversion."[6] The MPLA managed the transition to a multiparty democracy and market economy in a way that maintained its power while diversifying the modes, sites, and faces of its activity (private enterprise, charitable foundations, and international investments).[7] Media entities formerly tied to the state became independent of the MPLA party (at least in de jure terms); commercial and noncommercial bodies could now print papers and broadcast.

RNA lost its monopoly on broadcasting. The first commercial stations since independence opened in 1992: LAC, Luanda Antena Comercial (in Luanda); Rádio Morena (in Benguela); Rádio Comercial de Cabinda (in Cabinda); and Rádio 2000 (in Lubango).[8] In 1997 Rádio Ecclesia—the Catholic Broadcaster of Angola (belonging to the Episcopal Conference of Angolan and São Tomé)—began to broadcast for the first time in twenty years.[9] Rádio Despertar opened in 2006. Since 2008 Rádio Mais, another commercial station based in Luanda, has been broadcasting to Benguela, Huambo, and Huíla.[10] In the 2010s, new radio stations opened, particularly in Luanda: Rádio Kaíros (of the Methodist Church, 2012); Rádio UNIA (Independent University of Angola, 2013); Rádio Voz de Esperança (Tocoist Church, 2015); and Rádio Global FM (a community station).[11]

A 1999 Human Rights Watch report on the failures of the Lusaka Protocol pointed to the significance of radio listening in Angola. It noted that "the media—especially radio—is powerful in Angola and Angolans must be some of the most avid radio program listeners in the world (approximately 80% of 11 million inhabitants listen to radio)."[12] In the same report, Angolan NGO director Fernando Pacheco discussed the vibrancy of radio listening in rural areas, where he said nearly every village had two to three people who listened daily to the foreign services in Portuguese (BBC, VOA, RFI, and/or Antenna Africa from the SABC).[13] Many, he noted, also followed the RNA and UNITA stations, well aware of the interested position of each: "Radio has always been very popular and you often see people walking down the streets with radios pressed against their ears."[14] The advent of mobile phones, which are run on radio waves and nearly all of which offer radio-listening capabilities, has made radio more accessible.

Rádio Despertar opened in 2006. A commercial radio station, the owners of its founding business group, Socitel, are predominantly UNITA members. Mention Rádio Despertar and people think "UNITA."[15] Angolan politics continue to be bifurcated (though a third party, CASA-CE, is growing), even after the end of the civil war in 2002. This is a politicized peace in a land of feral capitalism.[16] Rádio Despertar struggles to stay afloat financially. Political intrigue, a murder, defections to RNA, and run-ins with the state have peppered its day-to-day operations.[17]

The Al Jazeera short documentary *Angola: Birth of a Movement*, shot three years before the arrests of seventeen civic activists in June 2015, offers close-ups of three activists: Luaty Beirão, Carbono Casimiro, and Mbanza Hamza.[18] Two of the three would be arrested in 2015 after various detentions and beatings across the years 2011 to 2015. In the documentary, these three are explicit about how media drive their intervention. The film dialogue opens with Hamza discussing the radio show *Zwela Angola*, which he hosted with Beirão on Rádio Despertar. Hamza places the show in relation to the MPLA's historic use of radio:

> For a long time, people have been asleep. They got used to hearing only one side of the story. Now it's time to open their eyes, pull back the veil, and show them what's really going on in the country. The liberation movements were born, they used radio to incite a revolutionary spirit in the people. We're going through the same process—trying to awaken people with our radio show.[19]

Like Angola's liberation movements, Hamza noted, he and Beirão used radio to "awaken" political consciousness. It was a story that ended badly. Badly for them and badly for the radiocractic claim that more radio stations equal more democracy. Their show was shut down by "superior orders,"[20] which came through the station manager, a UNITA member, and reeked of MPLA logic.[21]

Media scholars swept up in the enthusiasm of democratization on the African continent in the 1990s and early 2000s celebrated the accompanying de jure and de facto media liberalization. They equated more media outlets (newspapers, radio and TV stations, and content producers) with more democracy (greater access to information, an expanding public sphere, and space for individual voices). With reference to radio, this is what John Hartley dubbed "radiocracy."[22] The term "radiocracy" shorthands radio's capacity to expand the public sphere and promote participatory democracy. While the literature proliferates on community stations and their democratic ways, the polyphonic possibilities of rural broadcasting, and urban talk radio, a more recent literature points out constraints.[23]

When scholars embrace Western liberal notions of democracy and media-state configurations, it is not innocent. Transitions to democracy on the African continent accompanied structural adjustment programs that demanded state shrinkage and privatization. As Wendy Willems argues, the privatization of universities and the NGO-ization of civil society across the continent also shape media scholarship and journalism.[24] Whereas earlier African media scholars critiqued Western epistemologies inherent in modernization theory and ideas about the mass media that circulated routinely after World War II, privatization conditioned the work of journalists writing around and through the period of structural adjustment.[25] The sources of research financing were often NGOs and private (foreign) funders. It is perhaps no surprise, then, that academics as much as institutions have celebrated the relationship between radio and democracy. Yet this argument occludes the more sinister transformations at work in institutional structures, power relations, and what can be said.

The Angolan media ecology witnessed similar privatization, albeit less at the behest of Bretton Woods institutions.[26] The private, so-called "independent" media is largely owned by MPLA-affiliated investors.[27] As Péclard noted, the MPLA "arrogate[d] to itself the right to dictate the rules of the transition to peace and control the process of democratization."[28] The experience that Beirão and Hamza had at Rádio Despertar is a good example of the kind of squeeze on formal media space that takes place. Strikingly, that was at the hands of the official opposition. Other radio stations, like Rádio Ecclesia, struggle too. Closed for twenty years, taken over in 1977 and reopened in 1997, Ecclesia broadcast alternative sounds from the late 1990s through the early 2000s. Maka Angola, the website of investigative journalist Rafael Marques de Morais, described it as the "radio of the people."[29] The minister of communication accused it of being an antenna of terrorism in 2004. In the 2010s, journalists have complained of censorship, changes in leadership, an unclear legal status, and the ongoing refusal of the government to expand their broadcast range to the national level.[30]

If the trajectories of Rádio Despertar and Rádio Ecclesia are any indication, the state still holds radio close, even if the constitution secures broadcasting rights for bodies outside the state's formal control. The same is true for the printed press.[31] As of January 2007, 193 newspapers had been registered with the Ministry of Social Communication.[32] Very few actually function and circulate. Those that do are owned by people connected to the MPLA and former president dos Santos, even if editors are professional journalists.[33] Journalists remember the murder of their peer Ricardo de Melo in 1996. Melo's *Imparcial Fax* (1994–96) was initially circulated to subscribers by facsimile. It parsed politics, military secrets, and the behavior of the country's political leaders.[34] Former collaborating journalists refused to discuss the paper or Melo with fellow journalist and researcher Adebayo Vunge for his 2006 book.[35]

Media continue to change. Again just ahead of an election, the Angolan state responded with a package of new media laws in late January 2017. The package comprised five laws, including laws on the Exercise of Radio Broadcasting and Television, the Press Law, the Organic Law of the Regulatory Entity of Angolan Social Communication, and the Law on the Statute of the Journalist. Together these laws endeavor to control the media environment in new ways including what passes on social media.[36] The Union of Angolan Journalists, Human Rights Watch (HRW), and Maka Angola, among others, have criticized these laws for vague definitions of "public interest" and "defamation." Judges and ministers, representatives of the state, will adjudicate and define those concepts. The Ministry of Social Communication will administer the journalist's statute, not journalists. HRW notes the excessive cost of taxes imposed on new businesses wishing to open news agencies and radios (US$211,000 for the former, and US$452,000 for the latter).[37] This new set of laws is a direct response to the growth of private papers, independent journalists, and the work of activists who have been jamming the system with sounds old and new.

JAMMING

In June 2015, Angolan police arrested thirteen civic activists who had gathered to discuss Domingos da Cruz's manuscript "Ferramentas para derrubar um ditador e evitar nova ditadura: Filosofia politica de libertação para Angola" (Tools to remove a dictator and avoid a new dictatorship: Political philosophy of liberation for Angola). Da Cruz based his work on Gene Sharp's *From Dictatorship to Democracy*, a book that circulated underground for many years, that helped inspire nonviolent revolutions from Eastern Europe through Asia, and that activists and participants of the Occupy movement read closely. Arab Spring activists read it too.

Two days after the arrests of the thirteen activists, the police detained two more young men. A week later they arrested two young women before releasing them to home detention. The attorney general indicted all seventeen for "preparing acts pursuant to a coup d'etat." The Attorney general's office put forth the association of Sharp's book with the Arab Spring as evidence of sedition. The international media quickly dubbed the detainees the "Book Club."[38] This easy metaphor for political repression dressed their act of political engagement in the costume of Western liberal democracy.

The Portuguese editorial house Edições Tinta da China published a translated version of Sharp's book with a red flap folded around the cover that shouted: "Danger. In dictatorships, this book could put you in jail."[39] Portugal, erstwhile colonizer and a member of the democratic club only since 1974, now championed human rights, like free speech, in a repressive Angola through its NGOs and publishers.[40] But the problem was never the book. The book was a pretext: material evidence for the immaterial problem. The problem was the noise the activists generated. They were jamming the state signal, disrupting its frequency of power.

The activists are a diverse group, ranging in age from nineteen to thirty-three (in 2015). Some are primary and secondary school teachers, there is one university professor, some are students, two are journalists, three are hip-hop artists, and some are full-time civic activists employed at nongovernmental organizations (NGOs). They hail from diverse backgrounds. When the activists organized themselves in a study group, not a "book club," to think about how to better organize and protest President dos Santos and his oligarchy, the state lent an ear: using a tried-and-true PIDE tactic once used against MPLA militants, they sent an informer to infiltrate the group.

Of the activists arrested: Nito Alves, Luaty Beirão, Jeremias Benedito, Albano Evaristo Bingo, Osvaldo Caholo, Nuno Álvaro Dala, Sedrick de Carvalho, Rosa Conde, Domingos da Cruz, Inocencio de Brito, Nelson Dibango, Luarinda Gouveia, Mbanza Hamza, José Gomes Hata, Arante Kivuvu, Hitler Samussuko, and Fernando António Tomás "Nicola," a number had been involved in protesting President dos Santos's thirty-plus-year tenure since March 2011. Insisting on T-shirts and posters, in announcements, at concerts, and on handbills that "32 é muito" (32 is a lot—referring to the number of years President dos Santos had been in power at the first demonstration in 2011) and "Zé tira o pé" (Zé—the diminutive for José—quit blocking the way), the activists interrupted the state message and logics of working around and with the state that kept so many people quiet, if not silent.[41] They made noise publicly: organizing demonstrations calling for the president to step down, running civic education campaigns around the 2012 elections, producing "conscious" music, and hosting a radio show, among other activities.

It is important to rewind a bit and see how events unfolded to understand how these activists generated noise and why the symbols and models of liberal democracy fail to capture the power of their "jamming." In jamming you should hear both the jamming of the state frequency, the noise generation that Attali details, and the jamming of musical improvisation. It was their attempts to come together, to practice, to debate, to rehearse ideas and strategies, and to find the ways that worked (in jamming) that resulted in the state's striking back.

HISTORICAL NOISE: SCRAMBLING NATIONALIST PROPAGANDA

On 7 March 2011, an anonymous web-based announcement called for a demonstration in Luanda's Largo de Independência (Independence Plaza). Located just behind the RNA, the Largo has been the site of numerous rallies, always organized by the MPLA, whose headquarters abut the plaza. This demonstration is infamous for having been attended by more journalists and police than protesters and for having ended before it even began.[42] Attempting to pull the tide of the Arab Spring to Africa's southern shores, these activists called on their fellow citizens to show up, be seen and heard, and to tell the government that thirty-two years of rule by one president was enough. They broke the unwritten rules that censored public, collective protest and popular politics that have reigned since the tragic events of 1977.[43] But like Harold MacMillan's "Winds of Change" speech in the South African Parliament half a century earlier, reactionary political forces in southern Africa are sometimes mightier than the rhetorical invocation of natural change, at least in the short term.

"Agostinho Jonas Roberto dos Santos" signed the call to demonstrate on that March day—this was a pseudonym built from the names of the leaders of the country's founding anticolonial movements: Agostinho Neto (MPLA); Jonas Savimbi (UNITA); and Holden Roberto (UPA/FNLA)—with José Eduardo dos Santos's last name appended. It was signed in the name of a new movement: Movimento Revolucionário do Povo Lutador de Angola (Revolutionary Movement of the Struggling Angolan People). Unlike the nostalgia that pulses through so much of Angolan society and discourse, offering a subtle critique, this call did not look back. It created noise in the present by jamming the state message. The call used a direct, political language that scrambled the MPLA and state message, recasting elements of the past to intervene in the present.

Reformulating iconic names and dates, the announcement repossessed the patriotism of the leadership and called for broad change:

> The Revolutionary Movement of the Struggling People of Angola (MRPLA) is not a political party but a group of young Angolans that

> are looking for the development of a new social, political, and economic order in Angola. Although the rebellion against the government of José Eduardo dos Santos began some time ago, the MRPLA was founded on February 4, 2011, by Agostinho Jonas Roberto dos Santos who was inspired by the anti-government demonstrations in North Africa. The MRPLA is a movement of the Angolan people and it is not associated with any Angolan political party that has failed in the defense of the interests of the struggling people of Angola.[44]

With an acronym that differed by only one letter from that of the MPLA and claiming the same founding date (4 February) for its revolution, the MRPLA revolutionary movement sought to unsettle those in power and to grab the attention of fellow citizens.

Activists spread that call on social media, including Facebook, and circulated callouts for the demonstration via SMS. Doing research for this book in Luanda at the time, I received one such message on my flip phone. I received another warning me not to go. One cell company, Movícel, is a privatized subsidiary of the former Angola Telecom (the state telephone company); and the other, Unitel, is owned by dos Santos's eldest daughter, Isabel dos Santos.[45] Even though the state media were unavailable to them, the activists used the private phone system, a system that symbolizes state corruption, to broadcast their message.

Scholars and journalists and other activists have followed and documented the growth of this movement.[46] The state violence inflicted on protest participants and organizers, the arbitrary arrests, and the emergence of a larger social movement around them have garnered international attention and protest.[47]

The *revús* (short for revolutionaries), as Angolans refer to them, have received praise for their savvy use of social media. Their critics, on the other hand, dismiss them as disruptive and chaotic. In a 2013 interview on Portuguese television, then president dos Santos described them as "frustrated" and "failures at school."[48] They are noisy. They point out the regime's contradictions. Their call to protest and their criticism have been multimedia, employing analog and digital forms: a web announcement, Facebook posts, posters, phone trees, a radio program, and music videos. They refuse to form a political party. They do not have political ambitions.[49] They demand accountability and transparency. While their right to protest is guaranteed by the Angolan constitution, their ability to exercise that right is continually, maniacally, and nervously undermined.[50]

State violence, counterprotests to support the MPLA mounted every time activists organize an action, aggressive and dismissive MPLA rhetoric, and the curious handling of the arrest and trial of the seventeen activists that began in June 2015, as well as the ongoing legal wranglings with investigative reporter Rafael Marques de Morais, show just how unsettling, unnerving even, the activists have been to those in power. But little has been said about how this fits into historical practices of the media and mediation in Angola. Analysts are much more inclined to isolate these events as instances of human rights abuses and the corrupt state. The nervousness produced by these activists was palpable: police with water tanks and canine units meeting multigenerational, candle-holding peaceful protesters in 2015. Sometimes it was laughable: the jaunty, face-obscuring wig of one of the judges in the trial generated speculation that she did not want to be associated with the trial but did not have a choice.[51]

JAMMING THE JAMMERS

The activists are not the only critical voices. Rafael Marques de Morais has been an indefatigable force since the mid-1990s. His website, Maka Angola: In Defense of Democracy, against Corruption, uses the kimbundu noun *maka*, meaning "delicate, grave, or complex problem," to expose the abuse of power in Angola.[52] Marques de Morais refers to himself as a journalist and a human rights defender.[53] First and foremost, he is an investigative journalist. He is a thorn in the government's side, but he is too loud and too internationally known for them to seriously hurt or kill. Indeed, his continued existence and work demonstrate how he and the activists have shaped and opened the Angolan media ecosystem. Marques uses the Angolan legal system and the constitution against the state. Discontented bureaucrats and employees in the government ministries, institutes, the national oil company, and banks are among his informers.

Journalism has become a way to frame some of the methods and work of activists. Carbono Casimiro calls his work as a hip-hop artist "journalism," in this case, an alternative media source: "It is a way of communicating certain facts that aren't aired on the television or the radio."[54] In the Al Jazeera documentary, Casimiro walks through an area where *candongueiros* (informal collective taxis) pass. He and Hamza and others distribute free CDs of recorded music for drivers to play. This is an alternative mode of media circulation in the city. An alternative distribution system of communication, in Brecht's phrasing.[55] A transport system mobilized as a moving sound system by these new purveyors of information. Musicians, and activists, circumvent the state-owned media to send their ideas and sounds pulsing through urban space.[56] This practice is one kuduro artists developed to spread their music in the 1990s, outside the *circuito fechado* (closed circuit) of state- and DJ-monopolized media.[57] Kuduro artists, by and large, do not critique state power in contemporary Angola. Often, they have exchanged political complicity for state largesse.[58] But their do-it-yourself production and distribution techniques well serve a variety of ends.[59]

A handful of other artists—mostly hip-hop musicians but one or another kuduro artist—refer to their music as "conscious." Like the manifesto of the Movimento Revolucionário, this music is explicit in its subversion of the symbols, icons, and practices of MPLA power. It is never strident, often funny, always incisive. One example is Batida's "Cuca," featuring Ikonoklasta (aka Luaty Beirão) as a drunken war veteran who calls out the state's bread and circus (or beer and T-shirt) campaigns, which Buire describes as "a good example of the use of dominant cultural codes to defeat silent hegemony."[60]

What the activists, and Marques, have to say and the state's response to it—edgy, unnerved, disturbed by the noise—started to resonate with a growing number of the Angolan citizenry. In March 2011, many Angolans dismissed the call for a protest, calling it a secret service plot to measure public opinion before the election; not credible for being called anonymously; or jumping on the bandwagon of a political trend. The way the state overreacted to that event and the following protests, the murder of opposition CASA-CE member Manuel Ganga by state security, Marques's trial, the massacre of a religious sect in Huambo in April 2015, and the arrests of the 15+2 changed popular sentiment.[61] President dos Santos and senior members of the MPLA accused the activists of inciting war, claimed that Angola was not Tunisia (or Libya, or Egypt), and stated that anyone attending protests would be "neutralized" or "cleared away."[62] People began to take measure of the rhetoric and found it out of proportion to the activists' calls for peaceful action.[63] When the mothers of some of the jailed activists organized a protest and the state sent in police with attack dogs, more minds turned.[64]

The oil boom slowed dramatically by mid-2014.[65] A decade on from the peace, optimism waned and disappointment with the MPLA percolated as daily life became more expensive and precarious for most Angolans. What the state dismissed as noise came to sound like music to the ears of many Angolans.[66] Even as the MPLA party-state produces new iterations of powerful frequencies, sometimes privatized, sometimes as the law, Angolans find fresh grounds on which to protest and novel ways to assert their own interests. These rarely conform to neoliberal predictions or desires. At a moment when authoritarian regimes around the globe are on the rise, we should learn from these activists and the history of radio in authoritarian colonial and postcolonial regimes instead of being so sure we have lessons to teach them.[67]

Notes

INTRODUCTION

1. Frantz Fanon, *A Dying Colonialism*, trans. Haakon Chevalier (New York: Grove Press, 1965), 87.

2. Jay David Bolter and Richard A. Grusin, *Remediation: Understanding New Media* (Cambridge, MA: MIT Press, 2000). On Angola, see Delinda Collier's brilliant book *Repainting the Walls of Lunda: Information Colonialism and Angolan Art* (Minneapolis: University of Minnesota Press, 2016). I have written about the remediation of radio in two West African films: Marissa J. Moorman, "Radio Remediated: Sissako's *Life on Earth* and Sembène's *Moolaadé*," *Cinema Journal* 57, no. 1 (2017): 94–116.

3. Kirk Johnson, "As Low-Power Local Radio Rises, Tiny Voices Become a Collective Shout," 6 January 2018, *New York Times*, https://www.nytimes.com/2018/01/06/us/low-power-radio.html.

4. Bertolt Brecht, "Radio as a Means of Communication: A Talk on the Function of Radio," *Screen* 20, nos. 3–4 (December 1979): 25.

5. On the *Rachel Maddow Show*, 31 July 2018. See, too: Kara Swisher, "Zuckerburg: The Recode Interview," on Recode website, 18 July 2018, https://www.recode.net/2018/7/18/17575156/mark-zuckerberg-interview-facebook-recode-kara-swisher; and Evan Osnos, "Can Mark Zuckerburg Fix Facebook before It Breaks Democracy?," *New Yorker*, 17 September 2018.

6. Jonathan Auerbach and Russ Castronovo, eds., *The Oxford Handbook of Propaganda Studies* (New York: Oxford University Press, 2013).

7. Maria Teresa Prendergast and Thomas A. Prendergast, "The Invention of Propaganda: A Critical Commentary on and Translation of *Inscrutabili Divinae Providentiae Arcano*," in Auerbach and Castronovo, *Oxford Handbook of Propaganda Studies*, 19–27.

8. Priscilla Wald, "The 'Hidden Tyrant': Propaganda, Brainwashing, and Psycho-Politics in the Cold War Period," in Auerbach and Castronovo, *Oxford Handbook*

of Propaganda Studies, 115; and Auerbach and Castronovo, "Introduction: Thirteen Propositions about Propaganda," in Auerbach and Castronovo, *Oxford Handbook of Propaganda Studies*, 2.

9. Wald, "Hidden Tyrant," 113.

10. Nicholas J. Cull, "Roof for a House Divided: How U.S. Propaganda Evolved into Public Diplomacy," in Auerbach and Castronovo, *Oxford Handbook of Propaganda Studies*, 131–46; and Trysh Travis, "Books in the Cold War: Beyond 'Culture' and 'Information,'" in Auerbach and Castronovo, *Oxford Handbook of Propaganda Studies*, 180–200. For one powerful example of US Cold War cultural diplomacy in which the government used African American artists to promote the idea of US democracy even as Jim Crow continued, see Penny M. Von Eschen, *Satchmo Blows Up the World: Jazz Ambassadors Play the Cold War* (Cambridge, MA: Harvard University Press, 2004).

11. For work on how Salazar used Freyre's theory, see Cláudia Castelo, *"O modo português de estar no mundo": O luso-tropicalismo e a ideológica colonial portuguesa (1933–1961)* (Porto: Edições Afrontamento, 1998).

12. Eric Allina, *Slavery by Any Other Name: African Life under Company Rule in Colonial Mozambique* (Charlottesville: University of Virginia Press, 2012); Miguel Bandeira Jerónimo and António Costa Pinto, eds., *The Ends of European Colonial Empires: Cases and Comparisons* (New York: Palgrave Macmillan, 2015); Maria da Conceição Neto, "In Town and Out of Town: A Social History of Huambo (Angola), 1902–1961" (PhD thesis, SOAS, University of London, 2012); Manuela Ribeiro Sanches, Fernando Clara, João Ferreira Duarte, and Leonor Pires Martins, eds., *Europe in Black and White: Immigration, Race, and Identity in the "Old Continent"* (Chicago: University of Chicago Press, 2010); Boaventura de Sousa Santos, "Between Prospero and Caliban: Colonialism, Postcolonialism and Inter-Identity," *Luso-Brazilian Review* 39, no. 2 (2002): 9–43, among others.

13. Juanita Darling, "Radio and Revolution in El Salvador: Building a Community of Listeners in the Midst of Civil War, 1981–1992," *American Journalism* 24, no. 4 (2007): 67–93; Fanon, *Dying Colonialism*; Christopher Goscha, "Wiring Decolonization: Turning Technology against the Colonizer during the Indochina War, 1945–1954," *Comparative Studies in Society and History* 54, no. 4 (2012): 798–831; José Ignacio López Vigil, *Rebel Radio: The Story of El Salvador's Radio Venceremos*, abridged and trans. Mark Fried (Evanston, IL: Curbstone Books, 1995).

14. Like Ricardo Soares de Oliveira's *Magnificent and Beggar Land: Angola since the Civil War* (London: Hurst, 2015); and earlier work on the dynamic of war and state power, including Tony Hodges's *Angola: Anatomy of an Oil State* (first published as *Angola: From Afro-Stalinism to Petro-Diamond Capitalism*) (Bloomington: Indiana University Press, 2001); and Assis Malaquias, *Rebels and Robbers: Violence in Post-Colonial Angola* (Uppsala: Nordic Africa Institute, 2007).

15. Luise White, Stephan F. Miescher, and David William Cohen, eds, *African Words, African Voices: Critical Practices in Oral History* (Bloomington: Indiana University Press, 2001).

16. For more on this visit that marked the first formal talks between Cuba and the MPLA, see Edward George, *The Cuban Intervention in Angola, 1965–1991: From Che Guevara to Cuito Cuanavale* (New York: Frank Cass, 2005), 22.

17. On "the system," see Jon Schubert's *Working the System: A Political Ethnography of the New Angola* (Ithaca, NY: Cornell University Press, 2017).

18. The FBIS began in 1941 and continued until 1996. In 2005 the CIA's Open Source Center took up monitoring work: https://www.cia.gov/library/center-for-the-study-of-intelligence/csi-publications/books-and-monographs/foreign-broadcast-information-service/; and https://libraries.indiana.edu/foreign-broadcast-information-service-daily-reports-1941-1996.

19. Fernando Cavaleiro Ângelo, *Os flechas: A tropa secreta da PIDE/DGS na guerra de Angola* (Alfragide: Casa das Letras, 2016), 74.

20. Ângelo, *Os flechas*, 72–73, 75.

21. Renato Marques Pinto, "Os militares e as informações (em memória do General Pedro Cardoso)," in *Informações e segurança: Estudos em honra de General Pedro Cardoso*, ed. Adriano Moreira (Lisbon: Prefácio, 2004), 471–89; and Maria José Tiscar, *A PIDE no xadrez africano: Angola, Zaire, Guiné, Moçambique; Conversas com o Inspetor Fragoso Allas* (Lisbon: Edições Colibri, 2017), 41.

22. Frederick Cooper, *Decolonization and African Society: The Labor Question in French and British Africa* (Cambridge: Cambridge University Press, 1996); and Pinto, "Os militares e as informações," 473, specifically mentions the influence on the Portuguese military of the development of information systems and services during the Cold War.

23. Miguel Bandeira Jerónimo and António Costa Pinto, "A Modernizing Empire? Politics, Culture, and Economy in Portuguese Late Colonialism," in Jerónimo and Pinto, *Ends of European Colonial Empires*, 54.

24. Jerónimo and Pinto, "A Modernizing Empire?," 57.

25. Jerónimo and Pinto, "A Modernizing Empire?," 57.

26. Fernando Rosas, *História a história: África* (Lisbon: Tinta da China, 2018), 56.

27. Ângelo, *Os flechas*; and Rosas, *História a história*, 56–57.

28. Tiscar, *A PIDE no xadrez africano*, 40–41.

29. Ângelo, *Os flechas*, 34.

30. Pinto, "Os militares e as informações," 482; and Rosas, *História a história*, 55.

31. Ramiro Ladeiro Monteiro, "Subsídios para a história recente das informações em Portugal," in *Informações e segurança*, 459; and Pinto, "Os militares e as informações," 477.

32. Pinto, "Os militares e as informações," 477.

33. Pinto, "Os militares e as informações," 471–73.

34. Pinto, "Os militares e as informações," 480.

35. Ângelo, *Os flechas*, 71.

36. Pinto, "Os militares e as informações," 480.

37. Pinto, "Os militares e as informações," 480.

38. Rosas, *História a história*, 57–58; and see, too, Ângelo, *Os flechas*, 89.

39. Tiscar, *A PIDE no xadrez africano*, 39.

40. Rosas, *História a história*, 59–60.

41. Pinto, "Os militares e as informações," 480–81; and Ângelo, *Os flechas*, 79.

42. Ângelo, *Os flechas*, 79–80.

43. Tiscar, *A PIDE no xadrez africano*, 36; and Pinto, "Os militares e as informações," where this is the central theme and the reason Pinto eventually left SCCIA.

44. On the conflicts between SCCIA and PIDE/DGS in Angola, see Pinto, "Os militares e as informações," 477–79; and Ângelo, *Os flechas*, 71–79.

45. Ângelo, *Os flechas*, 80–83.

46. Ângelo, *Os flechas*, 93.

47. Ângelo, *Os flechas*, 90.

48. Ângelo, *Os flechas*, 91.

49. Ângelo, *Os flechas*, 91.

50. Aniceto Afonso and Carlos de Matos Gomes, *Alcora: O acordo secreto do colonialismo* (Lisbon: Objectiva, 2013), 159–204.

CHAPTER 1: SONIC COLONY

1. Sebastião Coelho, *Angola: História e estórias da informação* (Luanda: Executive Center, 1999), 123. The year was 1931 according to "Project" typed material from Rádio Nacional de Angola, undated but probably from 1990 or 1991 (private collection, Elisabete Pereira), and an eleventh-anniversary recognition in *Angola Radio* (Luanda: Largo da Sé), February 1942, 13, 15.

2. Miguel Gomes, dir., *Tabu*, DVD (Lisbon: O Som e a Fúria, 2012).

3. Mara Mills, "Deafness," in *Keywords in Sound Studies*, ed. David Novak and Matt Sakakeeny (Durham, NC: Duke University Press, 2015), 53. On the constraints of using deafness as a metaphor, see 52.

4. Thank you to Drew Thompson for suggesting I think about muteness.

5. Paul Landau, "Empires of the Visual: Photography and Colonial Administration in Africa," in *Images and Empires: Visuality in Colonial and Postcolonial Africa*, ed. Paul S. Landau and Deborah D. Kaspin (Berkeley: University of California Press, 2002), 141–71.

6. Sara Ahmed, "A Phenomenology of Whiteness," *Feminist Theory* 8, no. 2 (2007): 149–68. I thank Michelle Moyd for bringing Ahmed's work to my attention.

7. Jennifer Lynn Stoever analyzes "how listening operates as an organ of racial discernment" and the racialization of sound. Stoever, *The Sonic Color Line: Race and the Cultural Politics of Listening* (New York: New York University Press, 2016), 5.

8. On the Gold Coast, see Frederick Pratt, "'Ghana Muntie!': Broadcasting, Nation-Building, and Social Difference in the Gold Coast and Ghana, 1935–1985" (PhD diss., Indiana University, 2013); and on South Africa, see Thokozani N. Mhlambi, "Early Radio Broadcasting in South Africa: Culture, Modernity and Technology" (PhD diss., University of Cape Town, South African College of Music, 2015), esp. chap. 2, "Amateurs," and chap. 3, "Techno-Junkies," 11–24.

9. Brian Larkin, *Signal and Noise: Media, Infrastructure, and Urban Culture in Nigeria* (Durham, NC: Duke University Press, 2008), 57–58.

10. Debra Spitulnik, "Mobile Machines and Fluid Audiences: Rethinking Reception through Zambian Radio Culture," in *Media Worlds: Anthropology on New Terrain*, ed. Faye D. Ginsburg, Lila Abu-Lughod, and Brian Larkin (Berkeley: University of California Press, 2002), 227–54.

11. Debra Spitulnik, "Mediated Modernities: Encounters with the Electronic in Zambia," *Visual Anthropology Review* 14, no. 2 (1998–1999): 63–84.

12. Mhoze Chikowero, "Is Propaganda Modernity? Press and Radio for 'Africans' in Zambia, Zimbabwe, and Malawi during World War II and Its Aftermath," in *Modernization as Spectacle in Africa*, ed. Peter J. Bloom, Stephan F. Miescher, and Takyiwaa Manuh (Bloomington: Indiana University Press, 2014), 114.

13. Larkin, *Signal and Noise*, 59.

14. Peter J. Bloom, "Elocution, Englishness, and Empire: Film and Radio in Late Colonial Ghana," in Bloom, Miescher, and Manuh, *Modernization as Spectacle in*

Africa, 136–55; Chikowero, "Is Propaganda Modernity?," 112–35; David B. Coplan, "South African Radio in a Saucepan," in *Radio in Africa: Publics, Cultures, Communities*, ed. Liz Gunner, Dina Ligaga, and Dumisani Moyo (Johannesburg: Wits University Press, 2011), 134–48; Harri Englund, *Human Rights and African Airwaves: Mediating Equality on the Chichewa Radio* (Bloomington: Indiana University Press, 2011); Larkin, *Signal and Noise*, chap. 2; Sekibakiba Peter Lekgoathi, "Bantustan Identity, Censorship and Subversion on Northern Sotho Radio under Apartheid, 1960s–80s," in Gunner, Ligaga, and Moyo, *Radio in Africa*, 117–33; Pratt, "Ghana Muntie!"; Spitulnik, "Mobile Machines and Fluid Audiences"; and Spitulnik, "Mediated Modernities."

15. The state was also something of a latecomer to radio in South Africa where broadcasting started in 1924 and the state took over twelve years later. See Liz Gunner, "Wrestling with the Present, Beckoning to the Past: Contemporary Zulu Radio Drama," *Journal of Southern African Studies* 26, no. 2 (2000): 223.

16. Caroline Elkins and Susan Pedersen, "Introduction: Settler Colonialism; a Concept and Its Uses," in *Settler Colonialism in the Twentieth Century*, ed. Caroline Elkins and Susan Pedersen (New York: Routledge, 2005), 1–20. They lay out a typology that includes four groups: metropole (source of sovereignty); local administration (maintains order and authority); good-sized indigenous population; and a settler community. Settler colonialism is also characterized by inequalities and privilege that are codified in law, based on differences imagined between settlers and indigenous populations, and often racialized. See Elkins and Pedersen, "Introduction," 4. Aharon De Grassi points out the need to study the history of Angola as a settler colony and its implications for understanding Angola's geohistorical present. See De Grassi, "Provisional Reconstructions: Geo-Histories of Infrastructure and Agrarian Configuration in Malanje, Angola" (PhD diss., University of California, Berkeley, 2015), 80–81.

17. Cláudia Castelo, *Passagens para África: O povoamento de Angola e Moçambique com naturais da metrópole (1920–1974)* (Porto: Edições Afrontamento, 2007), 283.

18. Fernando Tavares Pimenta, *Angola, os brancos e a independência* (Lisbon: Edições Afrontamento, 2008), 60.

19. Maria da Conceição Neto, "In Town and Out of Town: A Social History of Huambo (Angola), 1902–1961" (PhD thesis, SOAS, University of London, 2012), 156.

20. Neto, "In Town and Out of Town," 156–57. Elizabeth Ceita Vera Cruz, *O estatuto do indigenato—Angola: A legalização da discriminação na colonização portuguesa* (Lisbon: Novo Imbondeiro, 2005).

21. Neto, "In Town and Out of Town," 158.

22. Castelo, *Passagens para África*, 285.

23. Castelo, *Passagens para África*; Jeanne Marie Penvenne, "Settling against the Tide: The Layered Contradictions of Twentieth-Century Portuguese Settlement in Mozambique," in Elkins and Pedersen, *Settler Colonialism in the Twentieth Century*, 79–94; Eric Morier-Genoud and Michel Cahen, eds., *Imperial Migrations: Colonial Communities and Diaspora in the Portuguese World* (New York: Palgrave Macmillan, 2012); and Eugénia Rodrigues, *A geração silenciada: A Liga Nacional Africana e a representação do branco em Angola na década de* 30 (Porto: Edições Afrontamento, 2003); on Portuguese settlers after empire, see Pamila Gupta, "'Going for a Sunday Drive': Angolan Decolonization, Learning Whiteness and the Portuguese Diaspora of South Africa," in *Narrating the Portuguese Diaspora: Piecing Things Together*, ed. Francisco Cota Fagundes, Irene Maria F. Blayer, Teresa F. A. Alves, and Teresa Cid (New York: Peter Lang, 2011), 135–52; and Gupta, "Decolonization and (Dis)Possession

in Lusophone Africa," in *Mobility Makes States: Migration and Power in Africa*, ed. Darshan Vigneswaran and Joel Quirk (Philadelphia: University of Pennsylvania Press, 2015), 169–93.

24. Luise White, *Unpopular Sovereignty: Rhodesian Independence and African Decolonization* (Chicago: University of Chicago Press, 2015), see, in particular, chap. 1.

25. Aniceto Afonso and Carlos de Matos Gomes, *Alcora: O acordo secreto do colonialismo* (Lisbon: Objectiva, 2013); and Maria Paula Meneses and Bruno Sena Martins, eds., *As guerras da libertação e os sonhos coloniais: Alianças secretas, mapas imaginados* (Lisbon: Almedina, 2013).

26. Gerald J. Bender, *Angola under the Portuguese: The Myth and the Reality* (Berkeley: University of California Press, 1972), 74 and 61–62; and Castelo, *Passagens para África*, 110–11, 202–4, 209. Investment in infrastructure (ports, roads, railroads, communications) increased with the First Development Plan (1953–1958), but Portuguese still immigrated overwhelmingly to Brazil and Europe throughout the 1950s–1970s. See Penvenne, "Settling against the Tide," 84–85.

27. Caroline Elkins and Susan Pedersen observe that tensions among four groups characterize twentieth-century settler colonialism: sovereign metropolitan power, a local/colonial administration, a majority indigenous population, and a minority settler community. See Elkins and Pedersen, "Introduction," 4. White's *Unpopular Sovereignty* underscores these tensions and the resultant messiness.

28. Miguel Bandeira Jerónimo and António Costa Pinto, "A Modernizing Empire? Politics, Culture, and Economy in Portuguese Late Colonialism," in *The Ends of European Colonial Empires: Cases and Comparisons*, ed. Jerónimo and Pinto (New York: Palgrave Macmillan, 2015), 56–57.

29. Bender, *Angola under the Portuguese*, 107.

30. The newly created Angolan Provincial Settlement Board—Junta do Povoamento Provincial—undertook this work. Bender, *Angola under the Portuguese*, 107–8.

31. Castelo, *Passagens para África*, 204.

32. Castelo, *Passagens para África*, 228.

33. Castelo, *Passagens para África*, 222–23.

34. Omar Ribeiro Thomaz, "Review of *Passagens para África*, by Cláudia Castelo," *Análise Social* 186 (January 2008): 186.

35. Castelo, *Passagens para África*, 204, 214.

36. Castelo, *Passagens para Àfrica*, 247–82. For Angola and other former Portuguese colonies, these aspects are evident in a recent spate of publications, predominantly journalistic and literary. See, for example, Dulce Maria Cardoso, *O retorno* (Lisbon: Tinta da China, 2012); Ana Sofia Fonseca, *Angola, terra prometida: A vida que os portugueses deixaram* (Lisbon: Esfera dos Livros, 2009); and Rita Garcia, *Luanda como ela era, 1960–1975: Histórias e memórias de uma cidade inesquecível* (Alfragide: Oficina do Livro, 2016).

37. David R. Roediger, *The Wages of Whiteness: Race and the Making of the American Working Class* (New York: Verso, 1991).

38. Castelo, *Passagens para África*; Dane K. Kennedy, *Islands of White: Settler Society and Culture in Kenya and Southern Rhodesia, 1890–1939* (Durham, NC: Duke University Press, 1987); and Brett L. Shadle, *The Souls of White Folk: White Settlers in Kenya, 1900–1920s* (Manchester: Manchester University Press, 2015).

In South Africa, the literature on the segregation period and apartheid highlights state welfare and promotion programs for Afrikaners aimed at improving their economic and educational attainments and giving them priority in state employment. See, for example, Deborah Posel, "The Case for a Welfare State: Poverty and the Politics of the Urban African Family in the 1930s and 1940s," in *South Africa's 1940s: Worlds of Possibilities*, ed. Saul Dubow and Alan Jeeves (Cape Town: Double Storey, 2005), 64–86; and Jeremy Seekings, "'Not a Single White Person Should Be Allowed to Go Under': *Swartgevaar* and the Origins of South Africa's Welfare State, 1924–1929," *Journal of African History* 48, no. 3 (2007): 375–94.

39. Ilídio do Amaral, *Aspectos do povoamento branco de Angola* (Lisbon: Junta de Investigações do Ultramar, 1960), 53; Castelo, *Passagens para África*, 221, 227–28. Benjamin Stora notes that this was also the case in Algeria. Stora, "The 'Southern' World of the *Pieds Noirs*: References to and Representations of Europeans in Colonial Algeria," in Elkins and Pedersen, *Settler Colonialism in the Twentieth Century*, 225–42. See, too, Maria da Conceição Neto's social history of Huambo, "In Town and Out of Town," for the history of a planned, white urban area inaugurated under Norton de Matos in 1912.

40. Castelo, *Passagens para África*, 203.

41. Castelo, *Passagens para África*, 266–67.

42. Didier Péclard, *Les incertitudes de la nation en Angola: Aux racines sociales de l'Unita* (Paris: Karthala, 2015), 79.

43. Cláudia Castelo, *"O modo português de estar no mundo": O luso-tropicalismo e a ideológica colonial portuguesa (1933–1961)* (Porto: Edições Afrontamento, 1998).

44. Fernando Tavares Pimenta, "Ideologia nacional dos brancos angolanos," in *Identidades, memórias e histórias, em terras africanas*, ed. Selma Pantoja (Brasilia: LGE, 2006), 172.

45. Stephen C. Lubkemann, "Unsettling the Metropole: Race and Settler Reincorporation in Postcolonial Portugal," in Elkins and Pedersen, *Settler Colonialism in the Twentieth Century*, 259. Lubkemann refers to this as "migrant colonialism" instead of "settler colonialism."

46. "Becoming White": A thanks to Jason Keith Fernandes for suggesting that I conceptualize this material in this manner.

47. Marcelo Bittencourt and Victor Andrade de Melo, "Esporte, economia, e política: O automobilismo em Angola (1957–1975)," *Topoi* 17, no. 32 (2016): 196–222.

48. Pimenta, *Angola, os brancos e a independência*, 345–426.

49. Ahmed, "Phenomenology of Whiteness."

50. Ahmed, "Phenomenology of Whiteness," 153–54.

51. This was key to white settler identity in Kenya and Rhodesia. See Shadle, *Souls of White Folk*, esp. chap. 2, pp. 26–57; and White, *Unpopular Sovereignty*, 35.

52. Gonçalo M. Tavares, *Breves notas sobre música* (Lisbon: Relógio D'Agua, 2015), 17.

53. Boaventura de Sousa Santos, "Between Prospero and Caliban: Colonialism, Postcolonialism and Inter-Identity," *Luso-Brazilian Review* 39, no. 2 (2002): 9–43. On whiteness, see pp. 28–33.

54. Santos, "Between Prospero and Caliban," 21.

55. Santos, "Between Prospero and Caliban," 21–24.

56. Jason Keith Fernandes, "A Goan Waltz around Post-Colonial Dogmas," unpublished paper from graduate workshop "Europe and the World: Between Colonialism and Globalization," Villa Vigoni, 18–22 June 2018. Cited with author's permission.

57. Pimenta, *Angola, os brancos e a independência*, 138. MPLA documents note that even though the colonial system used radio clubs indirectly to spread colonial ideology and political interests, the clubs sometimes took positions against the metropole. See Movimento Popular para a Libertação de Angola–Partido do Trabalho (hereafter MPLA-PT), "Teses e resoluções, 1° Congresso" (Luanda: Imprensa Nacional, 1978), 105.

58. Pimenta, "Ideologia nacional dos brancos angolanos," 177–78. Pimenta explains the limitations: categorized as Euro-Africans, they were excluded from upper posts in the public sector and the military; and the absence of universities in Angola made access to higher education difficult, as did state support of more recently arrived settlers. Pimenta cites a 1960 UN report on Angola that recognized a difference in rights and privileges between Portuguese born in Angola and those born in the metropole, though the greatest difference was between all of them and the largest part of the population: Africans classified as *indígenas*. This status informed the independence politics of whites and *mestiços* in the Organização Socialista de Angola (OSA) founded in Benguela. Pimenta, *Angola, os brancos e a independência*, 171.

59. Guilherme Mogas, interview by author, 11 May 2011, Luanda.

60. Pimenta, "A ditadura e a emergência do nacionalismo entre os brancos de Angola," in *Angola, os brancos e a indepêndencia*, 137–216.

61. The early works of Angolan writers like António Cardoso and Luandino Vieira are examples: Cardoso, *Baixa e musseques* (Havana: Ediciones Cubanas for the União de Escritores Angolanos, 1985); and José Luandino Vieira, *A cidade e a infância: Estórias*, 3rd ed. (Luanda: União de Escritores Angolanos, 1977).

62. Vieira, *A cidade e a infância*.

63. Pimenta, *Angola, os brancos e a independência*, 138–40, 162.

64. See Lila Ellen Gray, *Fado Resounding: Affective Politics and Urban Life* (Durham, NC: Duke University Press, 2013), chap. 2.

65. Gabriela Cruz, "The Suspended Voice of Amália Rodrigues," in *Music in Print and Beyond: Hildegard von Bingen to the Beatles*, ed. Craig A. Monson and Roberta Montemorra Marvin (Rochester, NY: University of Rochester Press, 2013), 180–99.

66. Cruz, "Suspended Voice."

67. Cruz, "Suspended Voice," 197.

68. John Durham Peters, "Helmholtz, Edison, and Sound History," in *Memory Bytes: History, Technology, and Digital Culture*, ed. Lauren Rabinovitz and Abraham Geil (Durham, NC: Duke University Press, 2004), 179.

69. Pimenta, *Angola, os brancos e independência*, 169. In particular Pimenta notes the connection between freemasonry (called Kuribeka in Angola—an Umbundu word), the local press, and the vindication of settlers' rights and, among some, independence.

70. Delinda Collier, *Repainting the Walls of Lunda: Information Colonialism and Angolan Art* (Minneapolis: University of Minnesota Press, 2016).

71. Castelo, *Passagens para África*, 252.

72. Carlos Alberto Medeiros, *A colonização das terras altas da Huíla (Angola)* (Lisbon: Centro de Estudos Geográficos, 1976), 629–30 (hereafter cited in text and notes as *CH*).

73. Medeiros points out that the first arrivals of Portuguese in the Huíla area date to the 1600s. Governor Sousa Countinho attempted white settlement in the mid-1700s, but the first successful, state-led settlement did not occur until 1857 (*CH*, 119, 125–28, 134–38). See also Bender, *Angola under the Portuguese*, 64.

74. Gervase Clarence-Smith, *The Third Portuguese Empire, 1825–1975: A Study in Economic Imperialism* (Manchester: Manchester University Press, 1985), 15. A frost hit crops hard. This group consisted of thirty Germans and twelve students from an orphanage. Later that year, more Germans and some Portuguese arrived (*CH*, 148–51). Brazilians from Pernambuco, not Portuguese, first settled in Moçamedes. Bender, *Angola under the Portuguese*, 72.

75. Medeiros notes that in 1883 the state ceded 3,966 hectares of land to Boers, at that point 325 in number (*CH*, 181). See, too, Clarence-Smith, *Third Portuguese Empire*, 44. Boers are responsible for transforming transportation in the region. Emmanuel Esteves, "As vias de comunicação e meio de transporte como factores de globalizacação, de estabilidade política e de transformação económica e social: Caso do Caminho-de-Ferro de Bengela (Benguela) (1889–1950)," in *Angola on the Move: Transport Routes, Communications and History/Angola em Movimento: Vias de Transporte, Comunicação e História*, ed. Beatrix Heintze and Achim von Oppen (Frankfurt am Main: Lembeck, 2008), 101.

76. Clarence-Smith, *Third Portuguese Empire*, 44–45.

77. Figures on the left remember that the white community of Lubango favored UNITA after 1974 and that the MPLA did not even have a delegation there. It was radio journalists, Leston Bandeira and Fernando Alves, who took over the radios and started broadcasting in favor of the MPLA. Interviews by author: Leston Bandeira, 17 June 2011, Lisbon; and Fernando Alves, 23 June 2015, Lisbon.

78. Roads connected the region to the central and northern regions, while the railroad was critical to transport to and from the coastal port city of Moçamedes.

79. See also Clarence-Smith, *Third Portuguese Empire*, 44.

80. Castelo's work shows that the white population in Sá de Bandeira nearly doubled, from 3,361 in 1940 to 6,201 in 1950. By 1970 it was estimated at 13,429. Castelo, *Passagens para África*, 222.

81. Franz Wilhem Heimer, *Decolonization in Angola*, 37.

82. See Fonseca, *Angola, terra prometida*, 163–80, on Sá de Bandeira.

83. Clarence-Smith, *Third Portuguese Empire*, 43.

84. Clarence-Smith, *Third Portuguese Empire*, 47.

85. This was not the case for Huambo/Nova Lisboa, which by 1960 had a preponderance of roads in the region and country. Road construction as a form of countersubversion grew significantly between 1961 and 1974, as Bender and others have noted. See Maria da Conceição Neto, "Nas malhas da rede: Aspectos do impacto económico e social do transporte rodoviário na região do Huambo c. 1920–c. 1960," in Heintze and von Oppen, *Angola on the Move*, 123–24.

86. Pimenta, *Angola, os brancos e a independência*, 110.

87. Pimenta, *Angola, os brancos e a independência*, 110.

88. Pimenta, *Angola, os brancos e a independência*, 294–304.

89. Pimenta, *Angola, os brancos e a independência*, 277.

90. Clarence-Smith, *Third Portuguese Empire*, 7, 49.

91. Castelo, *Passagens para Àfrica*, 221. While Santa Comba (the seat of the Cela *colonato*) was the most "European" city in terms of its population, of the five other cities Castelo lists with large white populations, only Sá de Bandeira qualifies as a midsize city.

92. Fonseca, *Angola, terra prometida*, 171.

93. Fonseca, *Angola, terra prometida*, 178.

94. Martin Heidegger, *The Question Concerning Technology and Other Essays*, trans. William Lovitt (New York: Garland, 1977), 16. Delinda Collier mentions this in discussing photography done by Diamang. Collier, *Repainting the Walls of Lunda*.

95. Ananya Jahanara Kabir's work emphasizes the links between modern technologies, the Black Atlantic, and social dances: "The Dancing Couple in Black Atlantic Space," in *Diasporic Women's Writing of the Black Atlantic: (En)Gendering Literature and Performance*, ed. Emilia María Durán-Almarza and Esther Álvarez López (New York: Routledge, 2013), 133–50; and Kabir, "Oceans, Cities, Islands: Sites and Routes of Afro-Diasporic Rhythm Cultures," *Atlantic Studies* 11, no. 1 (2014): 106–24. This is also the conceit of the Modern Moves research project she directed from 2013 to 2018, http://www.modernmoves.org.uk/.

96. Rádio e Televisão de Portugal (hereafter RTP), Área Museulógica e Documental, "Radio at the Service of Empire," *Angola Radio*, August 1941, 6.

97. RTP, Área Museulógica e Documental, *Angola Radio*, September 1941, 2.

98. RTP, Área Museulógica e Documental, *Angola Radio*, October 1941, 1.

99. Coelho, *Angola*, 240–41.

100. On commercial and agricultural associations, see Pimenta, *Angola, os brancos e a independência*, 94, 110, 294.

101. Interviews by author: Fernando Alves, 23 June 2015, Lisbon; Leston Bandeira,17 June 2011, Lisbon; and David Borges, 15 May 2008, Lisbon.

102. Interviews by author: Fernando Alves, 23 June 2015, Lisbon; Leston Bandeira, 17 June 2011, Lisbon; and David Borges, 15 May 2008, Lisbon.

103. See, for example, the website Angola Radio by Diamantino Pereira Monteiro, http://angolaradio.webs.com/, and his book *Rádio em Angola: Como eu a vivi* (Coimbra: Mar da Palavra, 2018). See also Fonseca, *Angola, terra prometida*; and José Maria Pinto de Almeida, *50 anos de rádio em Angola* (Casal de Cambra: Caleidoscópio, 2016).

104. A 1944 law made them a public utility. AOS/CO/UL 30B, pt. 1, ff. 75–77.

105. AOS/CO/UL 30B, pt. 1, ff. 75–77.

106. RTP, Área Museulógica e Documental, Emissora Nacional on "Radio Brazzaville e a Radiodifusão em Angola" [1959]. Caixa 163, "Emissora Nacional Técnica, Emissões Ultramarinas (1953–1973)," nine notes providing funding (almost negligible) from the governor-general for radio clubs.

107. Torre do Tombo, Arquivo Salazar, UL30B, caixa 733, pasta 1, p. 51.

108. Torre do Tombo, Arquivo Salazar, UL30B, caixa 733, pasta 1, pp. 67–68.

109. *Anuário Estatístico de Angola* (Luanda: Repartição Técnica de Estatística Geral, 1972), 99.

110. Depending on atmospheric interference, height of antennas, antenna gain, and antenna location, this would give them a 60–90 km range from the FM and medium wave but a greater, though less reliable, range from the shortwave.

111. Torre do Tombo, AOS/CO/UL/PC69/pasta 12/3–4 November 1964.

112. Torre do Tombo, AOS/CO/UL/PC69/pasta 12/3–4 November 1964. A 1969 letter from the radio broadcaster echoed these sentiments when it argued that its autonomy put it in an excellent position to broadcast counterpropaganda. See Instituto Diplomático, Lisbon: Gabinete de Negócios Políticos, Rádio Comercial de Angola, U-I-5.

113. Instituto Diplomático, Lisbon: Gabinete de Negócios Políticos, Rádio Comercial de Angola, U-I-5.

114. Fernando Alves, interview by author, 23 June 2015, Lisbon. José Oliveira noted the cutting-edge character of Angolan and particularly Lubango radio from the period and how these radio broadcasters later played key roles in remaking Portuguese television and radio after the military coup of 1974. José Oliveira, interview by author, 14 December 2015, Luanda. A point made by others as well: interviews by author: Ladislau Silva, 16 February 2011, Luanda; Leston Bandeira, 17 June 2011, Lisbon; Fernando Alves, 23 June 2015, Lisbon; Caro Proenca, 17 June 2013, Barreiro; and Paula Simons, 12 May 2011, Luanda.

115. Fernando Alves, interview by author, 23 June 2015, Lisbon.

116. Coelho, *Angola*, 125.

117. Coelho, *Angola*, 135n10.

118. Coelho, *Angola*, 194, 200–201n12.

119. Manuel Vinhas, a Portuguese businessman, started CUCA, among numerous other successful businesses in Angola. He took a distinct political approach and, for example, directly contacted the MPLA in hopes of negotiating and protecting his business interests. He believed Angolan independence was inevitable. The PIDE persecuted him and started a defamation campaign against him. Pimenta, *Angola, os brancos e a independência*, 303–4.

120. Facilitated by one of the directors of CUCA, engineer Albano Freitas, who was among the RC Luanda directorate. Note the overlap between commercial big men and radio management. Coelho, *Angola*, 197.

121. Marissa J. Moorman, *Intonations: A Social History of Music and Nation in Luanda, Angola, from 1945 to Recent Times* (Athens: Ohio University Press, 2008), chap. 4.

122. Coelho, *Angola*, 198; and on the recording industry, see 205–16.

123. Fernando Alves, interview by author, 23 June 2015, Lisbon.

124. Fernando Alves, interview by author, 23 June 2015, Lisbon.

125. Bittencourt and Melo, "Esporte, economia, e política," 216.

126. Bittencourt and Melo, "Esporte, economica e politica," 214.

127. On cultural models, see Marissa J. Moorman, "Putting on a *Pano* and Dancing Like Our Grandparents: Nation and Dress in Late Colonial Luanda," in *Fashioning Africa: Power and the Politics of Dress*, ed. Jean Allman (Bloomington: Indiana University Press, 2004), 84–103.

128. Bittencourt and Melo, "Esporte, economia, e política," 201, 204–8.

129. For a visual cultural example, see Zézé Gamboa's film *O grande kilapy* [The great swindle] (Lisbon: David and Golias, 2012), in which the young Joãozinho, a black Angolan playboy, participates in this fast-paced, largely white-dominated world. Also see this website: http://motorsportinangola.blogspot.fr/2009/12/motorsport-in-angola-old-times.html.

130. Bittencourt and Melo, "Esporte, economia, e política," 204–8; and on the rivalry between the Angolan and the Portuguese automobile clubs, see 208.

131. Bittencourt and Melo, "Esporte, economia, e política," 211.

132. See Coelho, *Angola*; and Moorman, *Intonations*, chap. 5, on the small-recording industry that took off in the early 1970s.

133. Bittencourt and Melo, "Esporte, economia, e política," 220.

134. Pimenta, *Angola, os brancos e a independência*, 216, 224.

135. Sócrates Dáskalos, *Um testemunho para a história de Angola: Do Huambo ao Huambo* (Lisbon: Vega, 2000), 82. The Movement for the National Liberation of

Angola (MLNA) started in Luanda in the mid-1950s. It advocated independence led by the black majority, whereas FUA asserted a Euro-African nationalism. See Pimenta, *Angola, os brancos e a independência*, 216.

136. Fernando Tavares Pimenta, *Angola no percurso de um nacionalista: Conversas com Adolfo Maria* (Lisbon: Edições Afrontamento, 2006), 63.

137. Pimenta, *Angola no percurso de um nacionalista*, 63.

138. Pimenta, *Angola no percurso de um nacionalista*, 66; and Dáskalos, *Um testemunho para a história de Angola*, 118.

139. Almeida, *50 anos de rádio em Angola*, 42–45, 60–63.

140. Bittencourt and Melo, "Esporte, economia, e política"; Pimenta, *Angola, no percurso de um nacionalista*; and Pimenta, "Ideologia nacional dos brancos angolanos."

141. Frederick Cooper and Ann Laura Stoler, eds., *Tensions of Empire: Colonial Cultures in a Bourgeois World* (Berkeley: University of California Press, 1997).

142. Nelson Ribeiro, "Broadcasting to the Portuguese Empire in Africa: Salazar's Singular Broadcasting Policy," *Critical Arts* 28, no. 6 (2014): 920–37.

143. Proença and Chaves remember that Rádio Clube de Angola's Humberto Mergulhão pushed the then director of the CTT to begin broadcasting. Caro Proença and Sara Chaves, interview by author, 17 June 2013, Barreiro.

144. Ribeiro, "Broadcasting to the Portuguese Empire in Africa," 925.

145. Ribeiro, "Broadcasting to the Portuguese Empire in Africa," 926.

146. Ribeiro, "Broadcasting to the Portuguese Empire in Africa," 928.

147. Ribeiro, "Broadcasting to the Portuguese Empire in Africa," 931.

148. James R. Brennan, "Radio Cairo and the Decolonization of East Africa, 1953–64," in *Making a World after Empire: The Bandung Moment and Its Political Afterlives*, ed. Christopher J. Lee (Athens: Ohio University Press, 2010), 173–95; Larkin, *Signal and Noise*, chap. 2; Graham Mytton, *Mass Communication in Africa* (London: Edward Arnold, 1983); Mytton, "From Saucepan to Dish: Radio and TV in Africa," in *African Broadcast Cultures: Radio in Transition*, ed. Richard Fardon and Graham Furniss (London: James Currey, 2000), 21–41; Spitulnik, "Mediated Modernities"; for an overview of radio on the continent after the fall of the Berlin Wall, see André-Jean Tudesq, *L'Afrique parle, L'Afrique écoute: Les radios en Afrique subsaharienne* (Paris: Karthala, 2002).

149. Coelho, *Angola*, 125. See the website Radiodifusão em Angola 1937/1975, http://angolaradio.webs.com/. The administrative part belonged to the CTT and the technical part belonged to Rádio Marconi and Rádio Clube de Angola. The colonial state, in this case through the governor-general's office, created official broadcasting that was dependent on local know-how.

150. Caro Proença and Sara Chaves, interview by author, 17 June 2013, Barreiro.

151. *Serão para Trabalhadores*, 3rd session of the 3rd series, EOA, from the RNA Historical Archives, recorded "Depoimento de Manuel António, Angola Livre guerrilheiro do MPLA."

152. Caro Proença and Sara Chaves, interview by author, 17 June 2013, Barreiro; Ribeiro, "Broadcasting to the Portuguese Empire in Africa," 931.

153. Caro Proença and Sara Chaves, interview by author, 17 June 2013, Barreiro. Coelho points to Humberto Mergulhão's and Natália Bispo's personal interests (they needed state pensions unavailable at the radio clubs) in promoting a national broadcaster. Coelho, *Angola*, 135–36n11.

154. See Ribeiro, "Broadcasting to the Portuguese Empire in Africa," 925.

155. Ribeiro, "Broadcasting to the Portuguese Empire in Africa," 933.

CHAPTER 2: GUERRILLA BROADCASTERS AND THE UNNERVED COLONIAL STATE IN ANGOLA

1. Joseph Sanchez Cervelló, "Caso Angola," in *Guerra colonial*, ed. Aniceto Afonso and Carlos de Matos Gomes (Lisbon: Editorial Notícias, 2000), 74; and Norrie MacQueen, *The Decolonization of Portuguese Africa: Metropolitan Revolution and the Dissolution of Empire* (New York: Longman, 1997), 35–36.

2. James M. Kushner, "African Liberation Broadcasting," *Journal of Broadcasting* 18, no. 3 (1974): 299.

3. Stephen R. Davis, "The African National Congress, Its Radio, Its Allies and Exile," *Journal of Southern African Studies* 35, no. 2 (2009): 349–73; Robert Heinze, "'It Recharged Our Batteries': Writing the History of the Voice of Namibia," *Journal of Namibian Studies* 15 (2014): 30; Sekibakiba Peter Lekgoathi, "The African National Congress's Radio Freedom and Its Audiences in Apartheid South Africa, 1963–1991," *Journal of African Media Studies* 2, no. 2 (2010): 141. The following papers were part of the Wits Workshop on Liberation Radios in Southern Africa, University of the Witwatersrand, February 2017, and cover ZANU, ZAPU, and FRELIMO broadcasting. Dumisani Moyo and Cris Chinaka, "Persuasion, Propaganda and Mass Mobilisation through Underground Radio in the Zimbabwe War of Liberation: A Comparative Study of Radio Voice of the People and Voice of the Revolution"; Eléusio dos Prazeres Viegas Filipe, "Voice of FRELIMO Insurrection against the Voice of Mozambique: Waging Propaganda War against Portuguese Colonialism and Building a National Consciousness, 1960s–1975"; and Alda Romão Saúte Saíde, "A *Voz da FRELIMO* and the Struggle for the Liberation of Mozambique, 1960s to 1970s."

4. John Mowitt, in *Radio: Essays in Bad Reception* (Berkeley: University of California Press, 2011), notes the construction of radio as a forgotten, understudied technology in radio studies.

5. Heinze, "It Recharged Our Batteries," 4. Work on radio in the United States also emphasizes the irony of relying on documentary sources to study sound: Susan J. Douglas, *Listening In: Radio and the American Imagination* (Minneapolis: University of Minnesota Press, 1999), 3, 9; Michele Hilmes, *Radio Voices: American Broadcasting, 1922–1952* (Minneapolis: University of Minnesota Press, 1997), xvi.

6. RNA has one recording of Angola Combatente from its Brazzaville and Lusaka years. The Associação Tchiweka de Documentação in Luanda holds fifteen to twenty transcripts of radio broadcasts from Brazzaville and Dar es Salaam, translated from Portuguese to English by Marga Holness. General Mbeto Traça, director of AC in Lusaka, and Guilherme Mogas, the second director of RNA, tried to recover materials from the exiled stations in Brazzaville and Lusaka but said they never managed to locate them and assume they have been lost. Interviews by author: Guilherme Mogas, 11 May 2011, Luanda; and General Mbeto Traça, 9 May 2011, Luanda.

7. Lekgoathi, "African National Congress's Radio Freedom," 144.

8. Namely, those from 1969 to 1973. For example, see PIDE/DGS Del. Angola, P Inf. 11.08.E, U.I. 1818, ff. 1–474 and 11.08.F, U.I. 1817, ff. 1–950. CHERET collected these transcripts and published them in their *Bulletin of Radio Listening* circulated to military and police with a certain security clearance. The PIDE transcribed broadcasts between 1964 and 1968 under different rules that did not mandate the disposal of the documents.

9. See, for example, the documents in PIDE/DGS, Del. Angola, P. Inf. 14.17.A, U.I. 2044, all of which are from this period.

10. See PT AHM/7B/13/4/273/40, Sit. Rep 339, September 1968, 6: the head of SCCIA calls this practice dangerous for weakening potential counterpropaganda.

11. Nancy Rose Hunt, "An Acoustic Register: Rape and Repetition in Congo," in *Imperial Debris: On Ruins and Ruination*, ed. Ann Laura Stoler (Durham, NC: Duke University Press, 2013), 39–66. On valuing delivery and affect and not just empirical evidence in oral histories, see Tom McCaskie, "Unspeakable Words, Unmasterable Feelings: Calamity and the Making of History in Asante," *Journal of African History* 59, no. 1 (2018): 3–5.

12. Mhoze Chikowero, "Is Propaganda Modernity? Press and Radio for 'Africans' in Zambia, Zimbabwe, and Malawi during World War II and Its Aftermath," in *Modernization as Spectacle in Africa*, ed. Peter J. Bloom, Stephan F. Miescher, and Takyiwaa Manuh (Bloomington: Indiana University Press, 2014), 113, 132.

13. Chikowero, "Is Propaganda Modernity?," 114.

14. Nancy Rose Hunt, *A Nervous State: Violence, Remedies, and Reverie in Colonial Congo* (Durham, NC: Duke University Press, 2016).

15. Dalila Cabrita Mateus, *A PIDE/DGS na guerra colonial, 1961–1974* (Lisbon: Terramar, 2004), 227.

16. Aharon De Grassi, "Rethinking the 1961 Baixa de Kassanje Revolt: Towards a Relational Geo-History of Angola," *Mulemba* 5, no. 10 (2015): 57–58.

17. De Grassi, "Rethinking the 1961 Baixa de Kassanje Revolt," 58.

18. Aida Freudenthal, "A Baixa de Cassanje: Algodão e revolta," *Revista Internacional de Estudos Africanos* 18–22 (1995–1999): 245–83. Freudenthal underscores that she is using colonial sources. She argues that the Portuguese colonial administration not only responded violently but then acted to conceal and downplay the events to cover their violence and obscure the causes of the revolt—namely, a pitiless forced-labor regime by the state-sanctioned concessionary Cotonang. See pp. 250–51 and 270–71.

19. Freudenthal, "A Baixa de Cassanje," 274.

20. Margarida Isabel Botelho Falcão Paredes, "Mulheres na luta armada em Angola: Memória, cultura e emancipação" (PhD diss., ISCTE–Instituto Universitário de Lisboa, 2014). See chapter 3, section 3.2.3, "A guerra da Maria"; and personal communication to author, 6 August 2018.

21. On the association between radio's invisible voices and the supernatural in the 1920s, radio's earliest days of broadcasting, see Jason Loviglio, *Radio's Intimate Public: Network Broadcasting and Mass-Mediated Democracy* (Minneapolis: University of Minnesota Press, 2005), xviii.

22. De Grassi, "Rethinking the 1961 Baixa de Kassanje Revolt," 53–133.

23. De Grassi, "Rethinking the 1961 Baixa de Kassanje Revolt," 117. De Grassi argues that "the discussion over the character of the revolt is also a proxy for other debates, including about moral claims to dignity and a share of national development" (65).

24. De Grassi, "Rethinking the 1961 Baixa de Kassanje Revolt," 91–94. Here De Grassi points beyond the movements of labor and road construction to villagization and urban reconstruction, a process he documents across the twentieth century and not just as a counterinsurgency strategy.

25. De Grassi, "Rethinking the 1961 Baixa de Kassanje Revolt," 58.

26. De Grassi, "Rethinking the 1961 Baixa de Kassanje Revolt," 60.

27. The arrests, trials, and imprisonment of nationalists in 1959 and 1960 demonstrate the Portuguese colonial state's disinterest in negotiated political rights. The state suppressed Angolan political organizing and refused to engage it. John A. Marcum, *The Anatomy of an Explosion (1950–1962)*, vol. 1 of *The Angolan Revolution* (Cambridge, MA: MIT Press, 1969), 33–34; and Douglas L. Wheeler and Réné Pélissier, *Angola* (New York: Praeger, 1971), 162–66.

28. De Grassi, "Rethinking the 1961 Baixa de Kassanje Revolt," 59; Freudenthal, "A Baixa de Cassanje," 252; Marcum, *Anatomy of an Explosion*, 125; and Wheeler and Pélissier, *Angola*, 174.

29. Freudenthal, "A Baixa de Cassanje," 276. Gerald J. Bender, *Angola under the Portuguese: The Myth and the Reality* (Berkeley: University of Califoria Press, 1978), 158, 165.

30. Marcum *Anatomy of an Explosion*, 129; and Wheeler and Pélissier, *Angola*, 175–76.

31. René Pélissier, *La colonie du Minotaure: Nationalismes et revoltes en Angola, 1926–1961* (Orgeval: Pélissier, 1978), chap. 14.

32. David Birmingham, *Frontline Nationalism in Angola and Mozambique* (Trenton, NJ: Africa World Press, 1992), 42.

33. For a compelling study of photography and political violence in this massacre, see Afonso Ramos, "Angola 1961, o horror das imagens," in *O império da visão: A fotografia no contexto colonial português (1860–1960)*, ed. Filipa Lowndes Vicente (Lisbon: Edições 70, 2014), 397–432.

34. The uncoordinated, violent response by the Portuguese state exacerbated tensions between white settlers and the state. Settlers pressured Prime Minister Salazar for reforms that would give them more representation and encourage foreign investment. See Fernando Tavares Pimenta, "O estado novo português e a reforma do estado colonial em Angola: O comportamento político das elites brancas (1961–1962)," *História* 33, no. 2 (2014): 250–72.

35. The war began in earnest in May 1961 when forces arrived from Portugal to occupy the north. See Aniceto Afonso and Carlos de Matos Gomes, eds., *Guerra colonial* (Lisbon: Editorial Notícias, 2000), 38–41.

36. Birmingham, *Frontline Nationalism*, 42.

37. The Trial of 50 marked the opening of the nationalist struggle. Police arrested fifty-six nationalist activists who were tried in three trials that named different organizations: ELA (Exército de Libertação de Angola); MIA (Movimento para a Independência de Angola); and MLA (Movimento para a Libertação de Angola), though numerous other small groups were involved. See Anabela Cunha, "'Processo dos 50': Memórias de luta clandestina pela independência de Angola," *Revista Angolana de Sociologia* 8 (2011): 87–96.

38. They were concerned with an Angolan-settled Portuguese citizen living in Brazzaville who was aligned with the Portuguese opposition. See chapter 3.

39. James R. Brennan, "Communications and Media in African History," in *The Oxford Handbook of Modern African History*, ed. John Parker and Richard Reid (Oxford: Oxford University Press, 2013), 501–2; and Heinze, "It Recharged Our Batteries," 30. A similar situation pertained in Algeria; see Frantz Fanon, *A Dying Colonialism*, trans. Haakon Chevalier (New York: Grove Press, 1965), 74.

40. This was the first consistent broadcasting. The MPLA broadcast briefly and inconsistently from Ghana in the early 1960s. Kushner, "African Liberation Broadcasting," 301–3.

41. The ANC's Radio Freedom likewise broadcast from Dar es Salaam and later Lusaka. Davis, "African National Congress," 379.

42. General Mbeto Traça, interview by author, 9 May 2011, Luanda.

43. Marcelo Bittencourt, *"Estamos juntos!" O MPLA e a luta anticolonial (1961–1974)*, vol. 1 (Luanda: Kilombelombe, 2008), 272; and Ilda Carreira, interview by author, 4 June 2008, Luanda.

44. See Don Barnett, "Liberation Support Movement Interview: Sixth Region Commander, Seta Likambuila, Movimento Popular de Libertação de Angola" (New York: Liberation Support Movement, 1971); Barnett, "Liberation Support Movement Interview: Member of MPLA Comité Director, Daniel Chipenda, Movimento Popular de Libertação de Angola" (New York: Liberation Support Movement, 1972); Augusta Conchiglia, *Guerra di popolo in Angola* (Rome: Lerici, 1969); and Basil Davidson, *In the Eye of the Storm: Angola's People* (New York: Doubleday, 1972). Support for radio journalism provided an easy route for foreign solidarity with the ANC, given the polarizing rhetoric of the Cold War. See Lekgoathi, "African National Congress's Radio Freedom," 142.

45. Heinze mentions support first from Radio Cairo in the 1950s, then Radio Ghana and Tanzania. Heinze, "It Recharged Our Batteries," 30. See also James R. Brennan, "Radio Cairo and the Decolonization of East Africa, 1953–64," in *Making a World after Empire: The Bandung Moment and Its Political Afterlives* (Athens: Ohio University Press, 2010), 173–95.

46. Institute of National Archives/Torre do Tombo, Lisbon, PIDE/DGS, Delegação de Angola, Processo de Informação 14.17.A, f. 915.

47. *Mário Bastos*, dir., *Independência* (Luanda: Geração 80, 2015).

48. Roberto indicated a Sr. Campos and the other was Hendrik Vaal Neto, recently the Angolan ambassador to Egypt and someone who left the FNLA for the MPLA in the 1980s. Holden Roberto, interview by author, 9 August 2005, Luanda.

49. All citations are from the following file: INA/TT, PIDE/DGS, Delegação de Angola, PInf. 11.08. E, U.I. 1818. Examples include: battlefront victories, ff. 927–29; addressing Portuguese soldiers, ff. 912–13 and "Portuguese oppressed by the Caetano regime," ff. 930–33; Roberto's visit to Bucharest, f. 800 and to China, ff. 927–29; and sending greetings to family and asking particular people to show up at the delegation headquarters, f. 911.

50. INA/TT, PIDE/DGS, Delegação de Angola, PInf. 11.08. E, U.I. 1818, ff. 912–13.

51. PT/AHM/FO/007/B/38, SSR 4- Angola, 1961970, 7/B 38/4 cx 360, pasta 18, rel tri APsic 4/68 1 Out a 31 dez 68, 1. See, too, PT/AHM/7B/13/4/273/40, Sit. Rep. 334, p. 2 of 3; and Sit. Rep. 339, 6 of 6.

52. Afonso and Gomes, *Guerra colonial*, 67; and John P. Cann, *Counterinsurgency in Africa: The Portuguese Way of War, 1961–1974* (Pinetown: Helion, 2012), chap. 3 and 6. These authors insist that more than in counterinsurgency wars in general, the Portuguese approach focused on soldiers and their work with local populations.

53. Mateus, *A PIDE/DGS na guerra colonial*, 376–78.

54. Mateus, *A PIDE/DGS na guerra colonial*, 381–82.

55. Mateus, *A PIDE/DGS na guerra colonial*, 187–93.

56. PT/AHM/7B/38/4/360, 1962 and PT/AHM/7B/13/4/306, the Orgânica da "Contra-Subversão," 15 April 1970.

57. This item received an F6 classification, which meant it was considered not trustworthy but deserving of further investigation. Torre do Tombo, PIDE/DGS, Delegação de Angola, PInf. 15.33.A, U.I. 2099, f. 275.

58. For an example from South Africa in the 1970s and 1980s, see Lekgoathi, "African National Congress's Radio Freedom," 143–45.

59. Bittencourt, "*Estamos juntos!*," 273.

60. Mário Bastos, dir., *Independência*, documentary (Luanda: Geração 80, 2015).

61. Portuguese state censorship was uneven. Scholars and journalists note more censorship in the metropole than in the African colonies (broadcaster Sebastião Coelho played songs banned in the metropole in Angola, for example). A censorship commission and a reading council followed print materials closely. These councils banned anything they deemed Marxist-Leninist, promoting Angolanidade (Angolanness), or possessing an Angolan perspective. They listened to radio but censored after the fact. See José Filipe Pinto, "A censura em Angola durante a guerra colonial," and interviews with Diamantino Pereira Monteiro and David Borges, in *O jornalismo Português e a guerra colonial*, ed. Silvia Torres (Lisbon: Guerra and Paz, 2016), 121–27, 173–94.

62. Mário Bastos, dir., *Independência*, documentary (Luanda: Geração 80, 2015). "Sachikwenda," a UNITA militant, was arrested and sent to Tarrafal Prison in 1969.

63. Mário Bastos, dir., *Independência*, documentary (Luanda: Geração 80, 2015).

64. Albina Assis, interview by author, 17 January 2002.

65. Lekgoathi, "African National Congress's Radio Freedom," 151.

66. Alberto Jaime, interview by author, 4 December 2001, Luanda.

67. Jardo Muekalia, *Angola—a Segunda revolução: Memórias da luta pela democracia* (Lisbon: Sextante, 2010), 16.

68. INA/TT, PIDE/DGS, Delegação de Angola, PInf. 14.17.A, U.I. 2044, f. 266. The author's tone and the mid-level security of the memo (C3 on a scale of A1-3–F1-3, most to least trustworthy) suggest that this was not a trained police officer but a neighborhood informer. Or, the tone might express frustration with the genuine inability of the PIDE to control the information in the territory and further evidence of the nervousness that independence in neighboring countries produced in the PIDE officers.

69. INA/TT, PIDE/DGS, Delegação de Angola, PInf. 14.17.A, U.I. 2044, f. 332, 14 July 1967.

70. INA/TT, PIDE/DGS, Delegação de Angola, PInf. 14.17.A, U.I. 2044, f. 101, 26 February 1968.

71. INA/TT, PIDE/DGS, Delegação de Angola, PInf. 14.17.A, U.I. 2044, f. 101.

72. See Marissa J. Moorman, *Intonations: A Social History of Music and Nation in Luanda, Angola, from 1945 to Recent Times* (Athens: Ohio University Press, 2008), 151. Albina Assis discusses how she and her female friends dissimulated listening by saying they were going to *bençon*, a gathering in the Catholic church, and a word that means "blessing."

73. Paul Apostolidis, *Stations of the Cross: Adorno and Christian Right Radio* (Durham, NC: Duke University Press, 2000); and John Durham Peters, *Speaking into the Air: A History of the Idea of Communication* (Chicago: University of Chicago Press, 2000), chap. 2. With reference to African broadcasting, see, for example, Dorothea E. Schulz, "Equivocal Resonances: Islamic Revival and Female Radio 'Preachers' in Urban Mali," in *Radio in Africa: Publics, Cultures, Communities*, ed. Liz Gunner, Dina Ligaga, and Dumisani Moyo (Johannesburg: Wits University Press, 2011), 63–80.

74. INA/TT, PIDE/DGS, Delegação de Angola, PInf. 14.17.A, U.I. 2044, ff. 279–81.

75. INA/TT, PIDE/DGS, Delegação de Angola, PInf. 14.17.A, U.I. 2044, f. 281.

76. Jean-Michel Mabeko Tali, *O MPLA perante si próprio*, vol. 2 of *Dissidências e poder de estado (1962–1977)* (Luanda: Nzila, 2001), 209–20; the chart on 219 shows

the composition of the Central Committee and the Political Bureau emerging from the Inter-Regional Conference of the MPLA in 1974. Five of the thirty-four members came from clandestine work or the prisons, while all others were active in the exiled guerrilla struggle.

77. INA/TT, Lisbon, PIDE/DGS, Delegação de Angola, Processo de Informação 15.28.A, f. 1076.

78. Alexander F. Toogood, "Portuguese Dependencies," in *Broadcasting in Africa: A Continental Survey of Radio and Television*, ed. Sydney W. Head (Philadelphia, PA: Temple University Press, 1974), 163.

79. See Lekgoathi, "African National Congress's Radio Freedom," 144, on group listening in South Africa. For examples of group listening in nonclandestine situations, see Harri Englund, *Human Rights and African Airwaves: Mediating Equality on Chichewa Radio* (Bloomington: Indiana University Press, 2011), 176–79; and Debra Spitulnik, "Documenting Radio Culture as Lived Experience: Reception Studies and the Mobile Machine in Zambia," in *African Broadcast Cultures: Radio in Transition*, ed. Richard Fardon and Graham Furniss (London: James Currey, 2000), 152–55.

80. INA/TT, Lisbon, PIDE/DGS, Delegação de Angola, Processo de Informação 14.17.A, f. 814.

81. INA/TT, PIDE/DGS, Delegação de Angola, PInf. 14.17.A, U.I. 2044.

82. Chikowero, "Is Propaganda Modernity?," 114.

83. By this he meant former *assimilados*. The state abolished the indigenato, which divided the African population into *assimilados* (assimilated or civilized) and *indígenas* (indigenous) after the uprisings of 1961, but class and cultural cleavages continued.

84. INA/TT, Lisbon, PIDE/DGS, Delegação de Angola, Processo de Informação 15.28.A, f. 1051. The military concerned itself with the impact of "enemy" propaganda on the white population. In PT/AHM/7B/38/4/360, 1962, pp. 1–4, the regional military commander for Angola reported to the military chief of staff in Lisbon that the white population, which is "theoretically and erroneously considered a priori pro-national, finds itself perplexed and disillusioned or already contaminated by the insinuating propaganda of the En."

85. David Borges remembered tuning in to *Angola Combatente* in Cunene but only becoming "politically conscious" after the Portuguese coup of 25 April 1974. See the interview with David Borges in Torres, *O jornalismo Português*, 187. José Oliveira remembers listening at home when serving in the colonial army in the late 1960s, with his mother leaning against his door at night at 7 p.m., nervously listening to hear if he was following the AC broadcast. José Oliveira, interview by author, 14 December 2015, Luanda.

86. INA/TT, PIDE/DGS, Delegação de Angola, PInf. 14.17.A, U.I. 2044, ff. 183–84, 234, and 191.

87. INA/TT, PIDE/DGS, Delegação de Angola, PInf. 14.17.A, U.I. 2044, f. 191, 1 December 1967.

88. INA/TT, PIDE/DGS, Delegação de Angola, PInf. 14.17.A, U.I. 2044, f. 113, 12 February 1968.

89. INA/TT, PIDE/DGS, Delegação de Angola, PInf. 14.17.A, U.I. 2044, f. 339. The transcriber noted that the soldiers did not speak Portuguese well, a suggestion they had been recruited from areas of thinnest Portuguese presence.

90. INA/TT, Lisbon, PIDE/DGS, Delegação de Angola, Processo de Informação 14.17.A, f. 885.

91. See Bender, *Angola under the Portuguese*. On Mozambique, see Allen F. Isaacman and Barbara S. Isaacman, *Dams, Displacement, and the Delusion of Development: Cahora Bassa and Its Legacies in Mozambique, 1965–2007* (Athens: Ohio University Press, 2013); and Isaacman and Isaacman, *Mozambique: From Colonialism to Revolution, 1900–1982* (Boulder, CO: Westview Press, 1993); and Allen F. Isaacman, *Cotton Is the Mother of Poverty: Peasants, Work, and Rural Struggle in Colonial Mozambique, 1938–1961* (Portsmouth, NH: Heinemann, 1996).

92. INA/TT, Lisbon, PIDE/DGS, Delegação de Angola, Processo de Informação 15.28.A, ff. 1065–67 and DGS report on Malange, 1964, f. 38.

93. All preceding citations from INA/TT, Lisbon, PIDE/DGS, Delegação de Angola, Processo de Informação 15.28.A, ff. 1065–67.

94. Brian Larkin, *Signal and Noise: Media, Infrastructure, and Urban Culture in Nigeria* (Durham, NC: Duke University Press, 2008), 40, 85; and Luise White, *Speaking with Vampires: Rumor and History in Colonial Africa* (Berkeley: University of California Press, 2000), 142–47. Both point out that these European representations of African responses tell us more about European projects and ideas than about Africans.

95. On the politics of independence and their relation to Congo, see Marcum, *Anatomy of an Explosion*, 60–64, 70–76, and 83–88. On Congolese independence, see Ch. Didier Gondola, *The History of the Congo* (Westport, CT: Greenwood Press, 2002); Georges Nzongola-Ntalaja, *The Congo from Leopold to Kabila: A People's History* (New York: Zed Books, 2002); and Crawford Young, *Politics in the Congo: Decolonization and Independence* (Princeton, NJ: Princeton University Press, 1965).

96. De Grassi, "Rethinking the 1961 Baixa de Kassanje Revolt," 57.

97. Hunt, *Nervous State*, 1.

98. Hunt, *Nervous State*, 8.

99. Chikowero, "Is Propaganda Modernity?," 114.

100. Hilmes, *Radio Voices*, 14.

101. Hilmes, *Radio Voices*, 14.

102. PT/AHM/FO/007/B/38, SSR 4, Angola 1962–1970, 7/B 38/4 cx 360, pasta 18, relatório trimestral APsic 4-68, 1 Out a 31 Dez 68, 2.

103. Larkin, *Signal and Noise*, 20.

104. Stefan Helmreich, "Transduction," in *Keywords in Sound Studies*, ed. David Novak and Matt Sakakeeny (Durham, NC: Duke University Press, 2015), 222–31.

105. Fanon, *Dying Colonialism*, 85.

106. Mowitt, *Radio*, 94.

107. Fanon, *Dying Colonialism*, 93–95.

108. Ingrid de Kok, "The Sound Engineer," in *Seasonal Fires: New and Selected Poems* (Houghton: Umuzi, 2006), 99–100.

109. De Kok, "Sound Engineer," 99.

110. De Kok, "Sound Engineer," 100.

111. Donald Meichenbaum, "Helping the Helpers," in *Trauma and Post-Traumatic Stress Disorder*, ed. Michael J. Scott and Stephen Palmer (London: Sage, 2000), 117–21; Aaron Reuben, "When PTSD Is Contagious," *Atlantic*, 14 December 2015, https://www.theatlantic.com/health/archive/2015/12/ptsd-secondary-trauma/420282/.

112. De Kok, "Sound Engineer," 99.

1. Pepetela, *A Geração da Utopia* (Lisbon: Dom Quixote, 1997), 154.

2. These are the great bulk of the contents of the one, five-hundred-page, file in the Torre do Tombo titled "Corangola: Comissão Coordenadora de Radiodifusão em Angola, PIDE Del Angola No. 15.33.A NT 2099." Other countries in the region also jammed clandestine liberation movement stations. See Dumisani Moyo, "Reincarnating Clandestine Radio in Post-Independent Zimbabwe," *Radio Journal* 8, no. 1 (2015): 33, who mentions how Ian Smith's regime jammed ZANU and ZAPU broadcasts with equipment from the Thornbill air base in Gweru.

3. Caroline Reuver-Cohen and William Jerman, eds., *Angola: Secret Government Documents on Counter-Subversion*, trans. Caroline Reuver-Cohen and William Jerman (Rome: IDOC, 1974), 89. Ian F. W. Beckett and John Pimlott note that during the counterinsurgency campaign, six times as much money went to road-building as to health and/or education. Beckett and Pimlott, *Counterinsurgency: Lessons from History* (Barnsley: Pen and Sword Books, 2011), 148.

4. Brian Larkin, *Signal and Noise: Media, Infrastructure, and Urban Culture in Nigeria* (Durham, NC: Duke University Press, 2008), discusses "noise" as the failures and missteps of technologies and how they are implemented.

5. The EN concerned itself with broadcasting critical of Portuguese rule from Brazzaville from the late 1950s and studied it closely in 1959. Emissora Nacional de Radiodifusão, "Rádio Brazzaville e a Radiodifusão em Angola," August 1959, RTP, Área Museulógica e Documental, caixa 69, Emissora Nacional Administração/Estudos, 1955–1972.

6. The state passed the same number of legislative actions on radio between 1930 and 1959 as between 1961 and 1969, to offer some measure of the new attention radio received. See *Angola: Resumo da Legislação, 1960–1969*.

7. The II Plano de Fomento Nacional, 1959–1961, passed in November 1958, merely indicated communications and transport in the overseas territories as one area of financial commitment. Under that rubric, "telecommunications" was the last of six listed targets of action. Diário Do Governo, I Série, no. 256, 25 November 1958, 1321; and Portaria 18272, Diário Do Governo, I Série, no. 40, 16 February 1961, 158.

8. Portaria 20416, 6 March 1964, Diário do Governo no. 56/1964, Série I de 6 March 1964.

9. Portaria 18357, Diário Do Governo, I Série, no. 71, 31 March 1961, 331. The commission thus answered directly to the governor-general of Angola, given that CITA had neither the resources nor the expertise to undertake the necessary changes.

10. Emissora Nacional de Radiodifusão, "Rádio Brazzaville e a Radiodifusão em Angola," August 1959, RTP, Área Museulógica e Documental, caixa 69, Emissora Nacional Administração/Estudos, 1955–1972; and Torre de Tombo, Arquivo Salazar, UL30B, caixa 733, pasta 1.

11. Decreto 43567, Diário Do Governo, I Série, no. 71, 27 March 1961, 334.

12. Portaria 19543, Diário Do Governo, I Série, no. 278, 4 December 1962, 1633.

13. Decreto 45237, Diário Do Governo, I Série, no. 212, 9 September 1963, 1454.

14. Untitled report, I Parte, "Problema Radiofónico Portugues," RTP, Área Museulógica e Documental, caixa 69, Emissora Nacional Administração/Estudos, 1955–1972, 1–2. Bívar likely wrote this report, though his name is not on it. It drew on

the Portuguese experience in the Spanish Civil War and the significant role of radio propaganda undertaken by the Portuguese Radio Club.

15. Untitled report, I Parte, "Problema Radiófonico Portugues," 2.

16. Untitled report, I Parte, "Problema Radiófonico Portugues," 2–4.

17. For example, Horst J. P. Bergmeier and Rainer E. Lotz, *Hitler's Airwaves: The Inside Story of Nazi Radio Broadcasting and Propaganda Swing* (New Haven, CT: Yale University Press, 1997).

18. Nelson Ribeiro, "Broadcasting to the Portuguese Empire in Africa: Salazar's Singular Broadcasting Policy," *Critical Arts* 28, no. 6 (2014): 921.

19. Ribeiro, "Broadcasting to the Portuguese Empire in Africa," 921; and Nelson Ribeiro, *A emissora nacional nos primeiros anos do estado novo, 1933–1945* (Lisbon: Quimera, 2005). Citation from Untitled report, I Parte, "Problema Radiofónico Portugues," 2.

20. Untitled report, II Parte, "Problemas Capitais de Emissões para Estrangeiro e África Negra," RTP, Área Museulógica e Documental, caixa 69, Emissora Nacional Administração/Estudos, 1955–1972, 1–2.

21. Untitled report, II Parte, "Problemas Capitais de Emissões para Estrangeiro e África Negra," 4.

22. Untitled report, II Parte, "Problemas Capitais de Emissões para Estrangeiro e África Negra," 4. Historian Júlio Mendes Lopes notes how the growth of radios (state and private) accompanied militarization, economic growth, and increased immigration between 1961 and 1974. See "A evolução da radiodifusão em Angola e a problemática da difusão da história da Àfrica" (PhD diss., Universidade de Agostinho Neto, Instituto Superior de Ciências de Educação, Luanda, 1997), 33.

23. Aniceto Afonso and Carlos de Matos Gomes, eds., *A conquista das almas: Cartazes e panlfetos da acção psicológica na guerra colonial* (Lisbon: Tinta da China, 2016), 18; and Pedro Aires Oliveira, "Live and Let Live: Britain and Portugal's Imperial Endgame (1945–75)," *Portuguese Studies* 29, no. 2 (2013): 186–208.

24. Miguel Bandeira Jerónimo and António Costa Pinto, "A Modernizing Empire? Politics, Culture, and Economy in Portuguese Late Colonialism," in *The Ends of European Colonial Empires: Cases and Comparisons*, ed. Jerónimo and Pinto (New York: Palgrave Macmillan, 2015), 51.

25. John P. Cann, *Counterinsurgency in Africa: The Portuguese Way of War, 1961–1974* (Pinetown: Helion, 2013), chap. 3 (hereafter cited in text as *CiA*).

26. They cited the US war plans and actions in Vietnam, but the emphasis on conventional war was less relevant.

27. Also see Reuver-Cohen and Jerman, *Angola*, 25.

28. Christopher R. Day and William S. Reno, "In Harm's Way: African Counter-Insurgency and Patronage Politics," *Civil Wars* 16, no. 2 (2014): 107–8.

29. The military Africanized troops in Angola particularly after a decline in the quality and enthusiasm of Portuguese recruits from 1966 on with the nationalist movements surge on the eastern front.

30. Fernando Tavares Pimenta, "O estado novo português e a reforma do estado colonial em Angola: O comportamento político das elites brancas (1961–1962)," *História* 33, no. 2 (2014): 250–72.

31. See also Afonso and Gomes, *A conquista das almas*, 19. The doctrine drew on studies of earlier counterinsurgency wars and on the first two years of fighting in Angola. From the English they drew on the idea of close military-civilian cooperation,

intelligence coordination, and small-unit operations. The French actions in Algeria inspired psychological operations. The guerrillas fought a low-technology war and cost was a primary concern for the Portuguese military. The study ran to five volumes.

32. See also Reuver-Cohen and Jerman, *Angola*.

33. Afonso and Gomes, *A conquista das almas*, 19–20. See PT/AHM/FO/007/B/38, SSR 4–Angola, 1962–1970, caixa 360. Various files focusing on the military's psychosocial action plans show a preoccupation with the MPLA's psychosocial action in particular (namely, its radio propaganda). See, too, AHM 2/2/14/89, 1969 Região Militar de Angola, Simpósio da contra-subversão: Conclusões finais e encerramento 1968/69, section V, 1.4; and PTAHM 7B 38 4 360 17, "RMA, QG, 3ª Repartição, ass: "Eficiência Operacional–Estado Moral das Tropas," 9 November 1962, 1–4.

34. Afonso and Gomes, *A conquista das almas*, 21.

35. Afonso and Gomes, *A conquista das almas*, 21–27. Psychological action was part of the army´s Second Division (Information) (2ª Repartição–Informações–of the Estado Maior do Exercito), and each local command in Angola, Guinea-Bissau, and Mozambique took responsibility for psychological actions in their territory. From 1968 to 1969 in Angola and Mozambique, the army set up separate Fifth Divisions for psychological action.

36. AHM 2/2/14/89, 1969 Região Militar de Angola, Doc: 1º relatório policopiado.

37. Irene Flunser Pimentel, *A história da PIDE* (Lisbon: Circulo de Leitores, 2011), 19.

38. AHM 2/2/14/89, 1969 Região Militar de Angola, relátorio, Ass: Simpósio da contra-subversao: Conclusões finais e encerramento 1968/69, I, 2.1.1.

39. The guitarist Matemona Sebastião had his mail for an electronics correspondence course opened repeatedly. Matemona Sebastião, interview by author, 27 February 2002, Luanda. The Luanda-based firm Luso-Holandesa proposed to take loudspeaker-outfitted trucks to rural areas to advertise small Phillips radios, turntables, electric razors, and other small items they sold. They promised to submit all texts to the PIDE prior to travel and to tune in only the state's *Voz de Angola* broadcast. PIDE Del Angola, No. 15.33.A NT 2099, ff. 415–16 and 437–39. Barcelo de Carvalho, known by his artistic name Bonga, requested funds to buy instruments and clothes for his folkloric dance group Kissueia and its fifteen members in 1963. PIDE/DGS, Del Angola, P.DInf, 10.05A, N.T. 2213, f. 46.

40. As per the example with Luso-Holandesa, whose request sat for more than eight months without response. PIDE Del Angola, No. 15.33.A NT 2099, ff. 415–16 and 437–39.

41. Ruy Llera Blanes, "Da confusão à ironia: Expectativas e legados da PIDE em Angola," *Análise Social* 206 (2013): 30–55.

42. Gerald J. Bender, *Angola under the Portuguese: The Myth and the Reality* (Berkeley: University of California Press, 1978), 160.

43. Bender, *Angola under the Portuguese*, 161–62, 164, 168–69, 178–79, 189, 193, and 195–96.

44. Bender, *Angola under the Portuguese*, 156–57.

45. Bender, *Angola under the Portuguese*, 157–58.

46. Daniel Branch, "Footprints in the Sand: British Colonial Counterinsurgency and the War in Iraq," *Politics and Society* 38, no. 1 (2010): 15–34. On the postindependence violence and political trouble sown by Africanization of Portuguese troops, see Carlos de Matos Gomes, "A africanização na guerra colonial e as suas sequelas tropas

locais: Os vilões nos ventos da história," in *As guerras da libertação e os sonhos coloniais: Alianças secretas, mapas imaginados*, ed. Maria Paula Meneses and Bruno Sena Martins (Lisbon: Almedina, 2013), 123–41.

47. Daniel Branch and Elisabeth J. Wood, "Revisiting Counterinsurgency," *Politics and Society* 38, no. 1 (2010): 3–14.

48. Bender, *Angola under the Portuguese*, 159, 170–71.

49. *Centro Emissor de Mulenvos* (Luanda: Centro Informação e Turismo de Angola, 1964), 1–15. Published by Luanda's Center for Information and Tourism, this document opens with the "terrorist" attacks of 1961 and the PRA as a swift response and the necessity of the transmission center to cover the rural areas not reached by the radio clubs and the EOA. See 3 and 6. Photos focus on the equipment: transmitters and towers. This is also the case in a photo album produced in 1969 when the center expanded. See TT/Arquivo Silva Cunha, cx 72, no. 518, mct 1.

50. Afonso and Gomes suggest that psychosocial action failed in Mozambique and Angola. Its only success was in helping Portuguese soldiers find meaning in their military service through bringing social assistance to Africans during the war when otherwise they delivered violence. Afonso and Gomes, *A conquista das almas*, 45.

51. Jerónimo and Pinto, "A Modernizing Empire?," 53–54.

52. Frederick Cooper, *Decolonization and African Society: The Labor Question in French and British Africa* (Cambridge: Cambridge University Press, 1996).

53. Jerónimo and Pinto, "Modernizing Empire?," 55–57.

54. Afonso and Gomes, *A conquista das almas*, 19.

55. Gervase Clarence-Smith, *The Third Portuguese Empire, 1825–1975: A Study in Economic Imperialism* (Manchester: Manchester University Press, 1985), 194.

56. Clarence-Smith, *Third Portuguese Empire*, 193.

57. Jerónimo and Pinto, "Modernizing Empire?," 63.

58. Jerónimo and Pinto, "Modernizing Empire?," 62–64. Jerónimo and Pinto point out key US support in the second and third plans, especially in the development of the petroleum industry; Jessica Marques Bonito, "Arquitectura moderna na África Lusófona: Recepção e difusão de ideias modernas em Angola e Moçambique" (master's thesis, Instituto Superior Técnico, Universidade Técnica de Lisbõa, 2011), 97; Allen F. Isaacman and Barbara S. Isaacman, *Dams, Displacement, and the Delusion of Development: Cahora Bassa and Its Legacies in Mozambique, 1965–2007* (Athens: Ohio University Press, 2013); and António Tomas, "The Making and Unmaking of a Modernist City" (chap. 1, p. 6, of unpublished manuscript, cited with the author's permission).

59. Jerónimo and Pinto, "Modernizing Empire?," 66–69.

60. Cláudio Castelo, *Passagens para África: O povoamento de Angola e Moçambique com naturais da metrópole (1920–1974)* (Porto: Edições Afrontamento, 2007), 143.

61. Ilídio do Amaral, "Contribuição para o conhecimento do fenómeno de urbanização em Angola," *Finisterra* 13, no. 25 (1978): 66; Bonito, "Arquitectura moderna na África Lusófona," 57; and Claudia Gastrow, "Negotiated Settlements: Housing and the Aesthetics of Citizenship in Luanda, Angola" (PhD diss., University of Chicago, 2014), chap. 1; this phenomenon was not lost on the colonial administration. In a five-page document from CITA in the military archives, the head of the Performance Services worried about the trouble that "illicit construction" caused vis-à-vis the proper authorization for recreational clubs in these burgeoning neighborhoods. AHM 2/2/169/19 CITA, 1–5.

62. Jerónimo and Pinto, "A Modernizing Empire?," 66.

63. Amaral, "Contribuição para o conhecimento," 70.

64. Jerónimo and Pinto, "A Modernizing Empire?," 60.

65. Ilídio do Amaral, *Luanda: Estudo de geografia urbana* (Lisbon: Memórias da Junta de Investigações do Ultramar, 1968); Filipa Fiúza and Ana Vaz Milheiro, "The Prenda District in Luanda: Building on Top of the Colonial City," in *Urban Planning in Lusophone African Countries*, ed. Carlos Nunes Silva (Surrey: Ashgate, 2015), 93–94; and Tomas, "Making and Unmaking of a Modernist City."

66. Tomas, "Making and Unmaking of a Modernist City," chap. 1, p. 19; and Ana Magalhães, *Moderno tropical: Arquitectura em Angola e Moçambique 1948–1975* (Lisbon: Tinta da China, 2009), 30–31.

67. Tomas, "Making and Unmaking of a Modernist City," chap. 1, pp. 11–14.

68. António Tomas argues that plans for urban modernization started as early as the regime of Norton de Matos as governor-general in the 1930s, when he promoted white settlement. In the early 1950s, a City Plan was commissioned from two urban planners, the Portuguese Moreira da Silva and the French Etienne de Gröer. They proposed radiocentric growth from a core based in Luanda's *baixa*. Five satellite cities would allow urban expansion (only two, Cacuaco and Viana, were ever realized). See Tomas, "Making and Unmaking of a Modernist City," 11.

69. Tomas, "Making and Unmaking of a Modernist City," 16.

70. Fernão Lopes Simões de Carvalho, interview by author, 11 June 2013, Queijas.

71. On the ideals and failure of Prenda, see Tomas, "Making and Unmaking of a Modernist City," chap. 1, pp. 23–24. See, too, Ana Tostões and Ana Maria Braga, "Unidade de vizinhança Prenda Luanda à luz da Carta de Atenas," in *Arquitectura em África Angola e Moçambique*, ed. Ana Tostões (Casal de Cambra: Caleidoscópio, 2008), 164–87.

72. Bonito, "Arquitectura moderna na África Lusófona," 115; Magalhães, *Moderno tropical*, 30–31. After finishing studies in architecture at the School of Fine Arts in Lisbon in 1955, he worked briefly at the Gabinete de Urbanização do Ultramar (Office of Overseas Urbanization) where he was disappointed with Portuguese urbanization practices and decided to go to Paris. He worked his way into Le Corbusier's atelier in Paris in 1957. At that time, Le Corbusier was in India working on Chandighar, and André Wogencsky ran his workshop. Eventually he enrolled at the Sorbonne´s Institute of Urbanization, to which he traveled back and forth from Luanda to complete his degree in 1965.

73. Simões de Carvalho, interview by author, 11 June 2013, Lisbon. He defines urban development as a combination of social, economic, and cultural development; see http://www.teoriadaarquitectura.com/artigo/fern-o-sim-es-de-carvalho-arquitectura-territ-rio-e-comunidade. See, too, Tostões and Braga, "Unidade de vizinhança," 166.

74. Jerónimo and Pinto, "Modernizing Empire?," 51–80.

75. Fernão Lopes Simões de Carvalho, interview by author, 11 June 2013, Queijas.

76. Simões de Carvalho complained that the building had been closed and air-conditioned, something both unnecessary and contrary to his original design. Fernão Simões de Carvalho, interview by author, 11 June 2013, Queijas. Among other features, Magalhães notes that the *brise soleil* denote some of the building's Corbusian elements. Malgalhões, *Moderno tropical*, 57–59.

77. António Tomás "Mutuality from Above: Urban Crisis, the State and the Work of *Comissões de Moradores* in Luanda," *Anthropology Southern Africa* 37, nos. 3–4 (2014):

175–86; Gastrow, "Negotiated Settlements"; in Mozambique this took the form of the Cahora Bassa dam, as Allen F. Isaacman and Barbara S. Isaacman have argued in *Dams, Displacement, and the Delusion of Development*. See Jeanne Penvenne's work on the racial organization of colonial life in Maputo, *African Workers and Colonial Racism: Mozambican Strategies and Struggles in Lourenço Marques, 1877–1962* (Portsmouth, NH: Heinemann, 1994).

78. Fernão Lopes Simões de Carvalho, interview by author, 11 June 2013, Queijas.

79. Afonso and Gomes, *A conquista das almas*, 21 and 25. Radio was not the only media used. Pamphlets and posters constituted the largest part of this campaign. In Mozambique, in Operation "Gordian Knot," loudspeakers were used for sonorous cajoling.

80. Reuver-Cohen and Jerman, *Angola*, 95.

81. David Borges, "A rádio de Angola estava muito mais desenvolvida do que a rádio do metropole," and Diamantino Pereira Monteiro, "A rádio de Angola era um prolongamento da rádio portuguesa," in *O jornalismo Português e a guerra colonial*, ed. Silvia Torres (Lisbon: Guerra and Paz, 2016), 185–88, and 174.

82. Cited on http://angolaradio.webs.com/ in the "Emissora Oficial" section. I was unable to locate the original document.

83. Torre do Tombo, SCCIA livro 221, Bases de estrutruração de um service de contrapropaganda, 31 December 1962, ff. 18–22.

84. This included fourteen pieces total signed by two different authors: V.S. and Maurício Ferreira in August and September 1967. The writer V.S. emphasized the following issues: the Plan's overreliance on technical infrastructure to the detriment of programming (having sidelined for years, for example, a proposal for a school for radio professionals put forward in 1962); the employment of metropolitan planners to solve Angolan problems they did not understand; and the need for a programming board that included local radio talent. See V.S., "Problemas da radiodifusão: Critério discutível," *ABC*, 28 August 1967, 5; "Problemas da radiodifusão: A E.O. largamente ultrapassada pelo Plano de Radiodifusão de Angola," *ABC*, 30 August 1967, 5; "Problemas da radiodifusão: Não estaremos na eminência de nos vermos transformados em 'aprendizes de feitiçeiro'?," *ABC*, 31 August 1967, 5; "Problemas da radiodifusão: A criação de um grupo orientador de programas afigura-se-nos a solução pela qual a E.O. deve enveredar," *ABC*, 2 September 1967, 5; "Problemas da radiodifusão: Orgânica e funcionamento de um conselho de programas," *ABC*, 3 September 1967, 5; "Problemas da radiodifusão: Impõe-se uma revisão de critérios eliminando previlégios que a E.N. não justifica," *ABC*, 4 September 1967, 4; "Problemas da radiodifusão: É necessário encarar com urgência a situação delicada das Emissoras de Cabinda e do Moxico," *ABC*, 5 September 1967, 5; "Problemas da radiodifusão: Entre um curso de teatro e uma escola de profissionais de rádio não será de optar por esta última solução?," IX, *ABC*, 6 September 1967, 5; "Problemas da radiodifusão: O caminho mais certo é o da igualdade de direitos entre os rádio clubes e as emissoras comerciais," *ABC*, 7 September 1967, 5.

85. Maurício Ferreira published four pieces as a "special to ABC": "Problemas da radiodifusão: O mais rápido 'veículo' de informação," *ABC*, 2 October 1967, 4; "Problemas da radiodifusão: O que é, de facto, a Emissora Oficial de Angola," *ABC*, 3 October 1967, 4; "Problemas da radiodifusão: Porque não a criação de um Instituto de Informação Pública?," *ABC*, 4 October 1967, 5; and "Problemas da radiodifusão: Temos de conceber uma rádio versátil e atraente," *ABC*, 6 October 1967, 4.

86. Ferreira, "Problemas de radiodifusão: O que é, de facto, a Emissora Oficial de Angola," *ABC*, 3 October 1967, 4.

87. V.S., "Problemas da radiodifusão: Impõe-se um revisão de critérios eliminando privilégios que a E.N. não justifica," *ABC*, 4 September 1967, 4.

88. Instituto Diplomático, "Memorial," from Rádio Comercial de Angola to the Gabinete de Negócios Políticos, Processo U-I-5, pasta "Rádio Comercial de Angola," 5 pp.

89. Instituto Diplomático, "Memorial," from Rádio Comercial de Angola to the Gabinete de Negócios Políticos, Processo U-I-5, pasta "Rádio Comercial de Angola," 5 pp.

90. AHM 2/2/14/89, 1969 Região Militar de Angola; and Reuver-Cohen and Jerman, *Angola*, 10.

91. Additional documents summarized actual discussions and decrees for a total of thirty-eight by the time they were finished. AHM 2/2/14/89, 1969 Região Militar de Angola; and Reuver-Cohen and Jerman, *Angola*, 11.

92. AHM 2/2/14/89, 1969 Região Militar de Angola, relátorio, Ass: Simpósio da contra-subversão: Conclusões finais e encerramento 1968/69.

93. AHM 2/2/14/89, 1969 Região Militar de Angola, relátorio, Ass: Simpósio da contra-subversão: conclusões finais e encerramento 1968/69, I, 2.1.6; and Reuver-Cohen and Jerman, *Angola*, 21.

94. AHM 2/2/14/89 1969, Região Militar de Angola, relátorio, Ass: Simpósio da contra-subversão: conclusões finais e encerramento 1968/69, I, 2.1.6.

95. Reuver-Cohen and Jerman, *Angola*, 34–35.

96. Reuver-Cohen and Jerman, *Angola*, 36–37. See James C. Scott, *Seeing Like a State: How Certain Schemes to Improve the Human Condition Have Failed* (New Haven, CT: Yale University Press, 1998), 1, on his early interest in how states try to impose a sedentary life on mobile peoples.

97. Reuver-Cohen and Jerman, *Angola*, 50. For comparative material on naming and identification in South Africa's pass system, see Keith Breckenridge, "Verwoerd's Bureau of Proof: Total Information in the Making of Apartheid," *History Workshop Journal* 59, no. 1, (2005): 83–108.

98. Reuver-Cohen and Jerman, *Angola*, 51.

99. Scott, *Seeing Like a State*.

100. Reuver-Cohen and Jerman, *Angola*, 93.

101. INA/TT, Lisbon, PIDE/DGS, Delegação de Angola, Processo de Informação 15.33.A, f. 57. This academic was Alexander Toogood, whose work I cited in chapter 2.

102. INA/TT, Lisbon, PIDE/DGS, Delegação de Angola, Processo de Informação 15.33.A, f. 37. Some administrators worried that the station's legal status needed to be resolved so that workers could either qualify as public servants or as unionizable workers of a private entity. Concern for employees appeared on other occasions: the commission likewise reported racism against station employees by bar owners in a downtown Luanda bar and looked at the case of an employee who was repeatedly assaulted and called out as a police informer. The counterinsurgency machine would not countenance chinks in the lusotropicalist armor.

103. An example from a letter to the Tchokwe language team from a soldier congratulating them on the program "Encontro a Noite." INA/TT, PIDE/DGS, Delegação de Angola, PInf. 15.33.A, U.I. 2099, f. 2.

104. One letter author, taken in to the PIDE for questioning, called out the announcers for assuming the position of "whites" and acting like they were better

than their fellow countrymen. INA/TT, Lisbon, PIDE/DGS, Delegação de Angola, Processo de Informação 15.33.A, ff. 138–39.

105. INA/TT, Lisbon, PIDE/DGS, Delegação de Angola, Processo de Informação 15.33.A, f. 417.

106. INA/TT, Lisbon, PIDE/DGS, Delegação de Angola, Processo de Informação 15.33.A, ff. 15, 353–54 for the Luena language program; ff. 347 and 348 for the Cuanhama language program.

107. INA/TT, Lisbon, PIDE/DGS, Delegação de Angola, Processo de Informação 15.33.A, f. 358.

108. INA/TT, Lisbon, PIDE/DGS, Delegação de Angola, Processo de Informação 15.33.A, f. 388 and 398.

109. INA/TT, Lisbon, PIDE/DGS, Delegação de Angola, Processo de Informação 15.33.A, ff. 376–78 and 382–84.

110. INA/TT, Lisbon, PIDE/DGS, Delegação de Angola, Processo de Informação 15.33.A, ff. 459 and 460.

111. INA/TT, Lisbon, PIDE/DGS, Delegação de Angola, Processo de Informação 15.33.A, ff. 315, 316, and 324.

112. Marissa J. Moorman, *Intonations: A Social History of Music and Nation in Luanda, Angola, from 1945 to Recent Times* (Athens: Ohio University Press, 2008), 156–57.

113. INA/TT, Lisbon, PIDE/DGS, Delegação de Angola, Processo de Informação 15.33.A, f465.

114. INA/TT, Lisbon, PIDE/DGS, Delegação de Angola, Processo de Informação 15.33.A, f. 417. See also f. 359 for the appeal of music to the youth, in particular.

115. PT/AHM/FO/007/B/38, SSR 4-Angola, 1962–1970, cx 360: Rel trimestral APsic 3/68, 1 jul a 30 set 68, 5; Rel trimestral APsic 3/68, 1 jul a 30 set 68, 5 & 12; and 7/B 38/4 cx 360, pasta 18, rel tri APsic 4/68 1 Out a 31 December 68, 24–25.

116. INA/TT, Lisbon, PIDE/DGS, Delegação de Angola, Processo de Informação 15.33.A, f. 359.

117. INA/TT, Lisbon, PIDE/DGS, Delegação de Angola, Processo de Informação 15.33.A, f. 417.

118. INA/TT, Lisbon, PIDE/DGS, Delegação de Angola, Processo de Informação 15.33.A, f. 126. For a similar one complaining of racism in the military camps and "Viva the MPLA!," see INA/TT, Lisbon, PIDE/DGS, Delegação de Angola, Processo de Informação 15.33.A, f. 213; and another criticizing the difference in salaries between Portuguese and Africans as well as housing, INA/TT, Lisbon, PIDE/DGS, Delegação de Angola, Processo de Informação 15.33.A, ff. 281–82, sent from Moçamedes in southern Angola in the name of the MPLA in 1969.

119. PIDE/DGS, Del Angola, Processo de Informação 11.08.F, U.I. 1817, f. 388.

120. INA/TT, Lisbon, PIDE/DGS, Delegação de Angola, Processo de Informação 15.33.A, f. 205.

121. INA/TT, Lisbon, PIDE/DGS, Delegação de Angola, Processo de Informação 15.33.A, f. 29.

122. INA/TT, Lisbon, PIDE/DGS, Delegação de Angola, Processo de Informação 15.33.A, f. 22. F. 146 for a positive assessment of the show sent from a French-speaking listener in Kinshasa.

123. INA/TT, Lisbon, PIDE/DGS, Delegação de Angola, Processo de Informação 15.33.A, f. 80.

124. INA/TT, Lisbon, PIDE/DGS, Delegação de Angola, Processo de Informação 15.33.A, f. 153.

125. "Elementos para a história da radiodifusão de Angola," *Electricidade* 87, 1970, 35; and Decreto no. 343/70, Diário de Governo, Série I, 20 July 1970, 962–76.

126. "Revisão da Orgânica de Contrasubversão 1974," AHM 2/2/127/4, No. 703/EC/74.

127. "Revisão da Orgânica de Contrasubversão 1974," AHM 2/2/127/4, No. 703/EC/74.

CHAPTER 4: NATIONALIZING RADIO

1. Marissa J. Moorman, *Intonations: A Social History of Music and Nation in Luanda, Angola, from 1945 to Recent Times* (Athens: Ohio University Press, 2008).

2. Nito Alves emphasized the importance of construction in a speech to a group of militants and the key role of the MPLA in this. He referred to the arrival of the nation as the coming "house" on the horizon in this speech in late October 1975 in which he denounced disunity and imperial, divisive forces like UNITA, Katangueses, and South Africans. "Nito Alves Junto dos Desalojados," 20 October 1975, RNA Historical Archives, recorded May 2011.

3. Jean-Michel Mabeko-Tali, *Guerrilhas e lutas sociais: O MPLA perante si próprio (1960–1977)* (Lisbon: Mercado de Letras Editores, 2018), 609.

4. Despacho no. 2/MINFA/75, Diário da República no. 25, 1a série, suplemento de 1975.

5. Manuel Ennes Ferreira, *A indústria em tempo de guerra (Angola, 1975–91)* (Lisbon: Edições Cosmos, 1999); Ferreira, "Nacionalização e confisco do capital português na indústria transformadora de Angola (1975–1990)," *Análise Social* 37, no. 162 (2002): 47–90.

6. Ricardo Soares de Oliveira, "Business Success, Angola-style: Postcolonial Politics and the Rise and Rise of Sonangol," *Journal of Modern African Studies* 45, no. 4 (2007): 595.

7. Oliveira, "Business Success," 598–603.

8. From the website of RNA, http://www.rna.ao/apresentacao/, accessed 27 November 2015.

9. Jean-Michel Mabeko Tali, *O MPLA perante si próprio*, vol. 2 of *Dissidências e o poder de estado (1962–1977)* (Luanda: Nzila Editora, 2001), 220–25.

10. The Alvor Accords created a transitional government with four-part administration: a Portuguese high commissioner (Brigadier Silva Cardoso) and the three presidents of the political parties (Agostinho Neto of the MPLA, Holden Roberto of the FNLA, and Jonas Savimbi of UNITA). They would hold elections to a constituent assembly by October 1975 and that assembly would draft a constitution. A new national army would be created with equal forces from the three movements, and twenty-four thousand Portuguese soldiers would remain in the territory until February 1976. See Norrie MacQueen, *The Decolonization of Portuguese Africa: Metropolitan Revolution and the Dissolution of Empire* (New York: Longman, 1997), 175–76.

11. The MFA took over the management of broadcasting at the EOA on 5 May 1974 due to what they deemed a delay in reporting the news. FBIS, Lisbon Domestic Service in Portuguese, 19h00 GMT, 5 May 1974.

12. Guilherme Mogas, interview by author, 11 May 2011, Luanda. This was not the case in Mozambique. There the director of FRELIMO's liberation radio became the

director of the new national broadcaster. See Alda Romão Saúte Saíde, "A *Voz da FRELIMO* and the Struggle for the Liberation of Mozambique, 1960s to 1970s," paper circulated at the Wits Workshop on Liberation Radios in Southern Africa, University of the Witswatersrand, February 2017.

13. Lara Pawson, *In the Name of the People: Angola's Forgotten Massacre* (New York: Tauris, 2014).

14. Tali, *Guerrilhas e lutas sociais*, 597–99 in particular, 594–603 more generally.

15. David Birmingham, "The Twenty-Seventh of May: An Historical Note on the Abortive 1977 'Coup' in Angola," *African Affairs* 77, no. 309 (October 1978): 554–64; Américo Cardoso Botelho, *Holocausto em Angola: Memórias de entre o cárcere e o cemitério* (Lisbon: Nova Vega, 2008); Miguel "Michel" Francisco, *Nuvem negra: O drama do 27 de Maio de 1977* (Lisbon: Clássica Editora, 2007); Dalila Cabrita Mateus and Álvaro Mateus, *Purga em Angola: Nito Alves, Sita Valles, Zé Van Dunem, o 27 de Maio de 1977* (Lisbon: Edições ASA, 2007); and Pawson, *In the Name of the People*.

16. Tali, *Guerrilhas e lutas sociais*, 595.

17. Tali, *Guerrilhas e lutas sociais*, 533, 603–12.

18. Interviews by author: Guilherme Mogas, 11 May 2011, Luanda; and Carlos Ferreira, 16 March 2011, Luanda. Elisabete Pereira, who was, at the time of independence, at *Voz de Angola*, said that of the fifteen employees of that program, twelve departed after the twenty-fifth of April coup in Portugal, leaving her and two of Angola's most highly regarded female vocalists: Lourdes Van-Dunem and Belita Palma. Elisabete Pereira, interview by author, 11 April 2011, Luanda.

19. Emanuel "Manino" Costa, interview by author, 25 May 2011, Luanda.

20. Emanuel "Manino" Costa, interview by author, 25 May 2011, Luanda; and Pascoal Mukuna, "Manino, ex-integrante do 'Kudibangela' e sobrevivente da purga," *Semanário Angolense*, 29 May 2010, 7–11.

21. João Filipe Martins, "Linhas mestres para informação Angolana," 20 December 1975, Arquivo Histórico, Rádio Nacional de Angola.

22. Emanuel "Manino" Costa, interview by author, 25 May 2011, Luanda; and Pascoal Mukuna, "Manino, ex-integrante do 'Kudibangela' e sobrevivente da purga," *Semanário Angolense*, 29 May 2010, 7–11.

23. Tali, *O MPLA perante si próprio*, 74, 71–72.

24. Tali, *O MPLA perante si próprio*, 74; and Michael Wolfers and Jane Bergerol, *Angola in the Frontline* (London: Zed Books, 1983), 75–76. In *Guerrilhas e lutas sociais*, Tali also mentions the FAPLA radio program *Povo em Armas* (576).

25. Wolfers and Bergerol, *Angola in the Frontline*, 75.

26. Pawson, *In the Name of the People*, 204; and Tatiana Pereira Leite Pinto, "Etnicidade, racismo e luta em Angola: As questões étnicas e raciais na luta de libertação e no governo de Agostinho Neto" (BA diss., Universidade Federal Fluminense, 2008).

27. Pinto, "Etnicidade, racismo e luta," 44–57.

28. Pawson, *In the Name of the People*, 101.

29. Tali, *Guerrilhas e lutas sociais*, 600.

30. Francisco, *Nuvem negra*, 35; Mateus and Mateus, *Purga em Angola*, 84–85; and Pawson, *In the Name of the People*, 181.

31. Pawson, *In the Name of the People*, 58–61. This account is the basis for the narrative in Wolfers and Bergerol, *Angola in the Frontline*, 90–95.

32. Mateus and Mateus, *Purga em Angola*, 88–89.

33. Ricardo Soares de Oliveira, *Magnificent and Beggar Land: Angola since the Civil War* (London: Hurst, 2015), 96. See, too, Jon Schubert, *Working the System: A Political Ethnography of the New Angola* (Ithaca, NY: Cornell University Press, 2017); Tali, *O MPLA perante si próprio*, 218–24; and, for a fictional rendering, Manuel Rui, *Crónica de um Mujimbo* (Lisbon: Cotovia, 1999).

34. Ferreira, "Nacionalização e confisco do capital português," 49; and Ferreira, *A indústria em tempo de guerra*, 181.

35. Ferreira, *A indústria em tempo de guerra*, 180.

36. Ferreira, *A indústria em tempo de guerra*, 181–82.

37. In his most recent work, Tali asks whether or not this was a direct result of the pressures put on the party by the arguments of Nito and Van-Dunem and their followers who wanted a Soviet-style Marxist-Leninist system. See Tali, *Guerrilhas e lutas sociais*, 613–14.

38. Ferreira, *A indústria em tempo de guerra*, 16; and MPLA-PT, "Teses e resoluções, 1° Congresso" (Luanda: Imprensa Nacional, 1978).

39. Ferreira, *A indústria em tempo de guerra*, 201.

40. See Ferreira, *A indústria em tempo de guerra*, 19, 27–29, on particulars of the policy outcomes of the First MPLA Party Congress.

41. Ferreira, *A indústria em tempo de guerra*, 205.

42. Sebastião Coelho, *Angola: História e estórias da informação* (Luanda: Executive Center, 1999), 130, and esp. 134n7. Coelho explains his official opposition to this form of organization submitted in a report he wrote as a hired consultant for the Ministry of Information to then information minister João Filipe Martins. One of the problems of monopoly, Coelho argued, was that a singular hold on all national broadcasting meant that if it fell into the wrong hands, the broadcaster would have access to all the transmitters in the country. Despite Martin's "full confidence" in the then director of the radio (who was later killed as a coup plotter), Coelho noted that though the *Kudibangela* group did not take advantage of it, for an hour and a half they "had the country in their hands."

43. "Programa 1° Congresso do MPLA, 10 Emissão," 16 October 1977, Arquivo Histórico, Rádio Nacional de Angola. Ironically, this was recorded over fado, characteristically Portuguese music.

44. See Pawson, *In the Name of the People*, esp. chapter 20, "Metamorphoses of the Enemy," about Costa Andrade "Ndunduma." His editorials encouraged people to "strike while the iron is hot"; and Tali, *O MPLA perante si próprio*, 221. These announcements enter in José Eduardo Agualusa's novel *A teoria geral do esquecimento* (Lisbon: Dom Quixote, 2012), 71: "An irate voice vociferated from the radio: *We must find them, tie them up, and shoot them!*"

45. Cited in MPLA-PT, "Teses e resoluções, 1° Congresso," 109.

46. Palavra do Neto na Rádio Nacional de Angola, 5 October 1977, Arquivo Histórico, Rádio Nacional de Angola.

47. See this piece in *Jornal de Angola* that celebrates RNA's thirty-fifth anniversary in 2012: http://jornaldeangola.sapo.ao/sociedade/mpla_sauda_a_direccao_e_trabalhadores_da_rna.

48. José Patrício, interview by author, 13 June 2012, Luanda. Patrício spent nearly a decade at RNA. He worked in the presidency and was later ambassador to the United States and to Portugal. Carlos Ferreira emphasized the excellent education in writing and speaking that young journalists received in the early days of RNA as important to

the ascension of these figures. Carlos Monteiro Ferreira, interview by author, 16 March 2011, Luanda.

49. M. Anne Pitcher, *Transforming Mozambique: The Politics of Privatization, 1975–2000* (Cambridge: Cambridge University Press, 2002), 85–86.

50. Pitcher, *Transforming Mozambique*, esp. chap. 2.

51. Palavra do Neto na Rádio Nacional de Angola, 5 October 1977, Arquivo Histórico, Rádio Nacional de Angola.

52. Fernando Alves, interview by author, 23 June 2015, Lisbon.

53. Interviews by author: Carlos Monteiro Ferreira, 16 March 2011, Luanda; and Bruno Lara, 17 April 2011, Luanda.

54. Angola Combatente broadcasts contained a vein of party ideology and reading of Portuguese colonialism and of military events that inflected it. See Marcelo Bittencourt, *"Estamos juntos!": O MPLA e a luta antiocolonial, 1961–1974*, 2 vols. (Luanda: Kilombelombe, 2008).

55. David Birmingham, *A Short History of Modern Angola* (London: Hurst, 2015), 81, 93–94.

56. On the importance of MDM to the party's revolutionary project, see MPLA-PT, "Teses e resoluções, 1° Congresso," 106–7, 125.

57. RNA 1990 Annual Report.

58. Jeffrey Herbst, *States and Power in Africa: Comparative Lessons in Authority and Control* (Princeton, NJ: Princeton University Press, 2014).

59. See Neto, "Nas malhas da rede: Aspectos do impacto económico e social do transporte rodoviário na região do Huambo c. 1920–c. 1960," in *Angola on the Move: Transport Routes, Communications and History/Angola em Movimento: Vias de Transporte, Comunicação e História*, ed. Beatrix Heintze and Achim von Oppen (Frankfurt am Main: Lembeck, 2008), 117, for a beautiful analysis of the significance of roads and the weight of the word *circulate* in contemporary, war-torn Angola.

60. Oliveira, "Business Success," 599.

61. Oliveira, "Business Success," 601.

62. See Ferreira, *A indústria em tempo de guerra*, 36–37, on the approval of the MPLA needed for appointments. José Oliveira, interview by author, 14 December 2015, Luanda, on the specificity of this to RNA.

63. José Oliveira, interview by author, 14 December 2015, Luanda. These other institutions often complained of not having their speeches aired in their entirety (generally reserved for President Neto and Lúcio Lara). Protection also extended to individuals. Oliveira tells the story of not being sent to prison in 1980 when Manuel Pedro Pacavira wanted to take over ANGOP, the news service, and accused him and João Melo of being spies. They were subsequently suspended but not imprisoned. Eventually (after two to three years of suspension), both were "rehabilitated." Melo became director of *Jornal de Angola*, and Oliveira was sent to reenergize the magazine *Novembro* as director of information (under Roberto de Almeida, secretary for information on the BP).

64. Interviews by author: Guilherme Mogas, 11 and 24 May 2011, Luanda; and José Oliveira, 14 December 2015, Luanda; Rui Amaro, conversation with author, 24 May 2011, Luanda; Julio Mendes Lopes, "A evolução da radiodifusão em Angola em Angola e a problemática da difusão da história da Àfrica" (PhD diss., Universidade de Agostinho Neto, Instituto Superior de Ciências de Educação, Luanda, 1997). Lopes was also a radio journalist.

65. Interviews by author: Guilherme Mogas, 11 and 24 May 2011, Luanda; Carlos Monteiro Ferreira, 16 March 2011, Luanda; and José Oliveira, 14 December 2015,

Luanda. See, too, "Carta louvar à Rádio Nacional de Angola" do Gabinete do Vice Ministro de Plano para o Director da Rádio Nacional de Angola, 1987. RNA documents.

66. Unauthored document, Centro de Documentação, RNA, notes, 6 December 2010, Luanda.

67. Interviews by author: Bruno Lara, 17 April 2011, Luanda; Carlos Monteiro Ferreira, 16 March 2011, Luanda; Guilherme Mogas, 11 and 24 May 2011, Luanda.

68. Ferreira, *A indústria em tempo de guerra*, 37; and Tali, *O MPLA perante si próprio*, 218–24; and Jean-Michel Mabeko, "Jeunesse en armes: Naissance et mort d'un rêve juvénile de démocratie populaire en Angola en 1974–1977," in *L'Afrique des générations: Entre tensiones et négociations* (Paris: Karthala, 2012), 301–58.

69. Ferreira, *A indústria em tempo de guerra*, 25.

70. Ferreira, *A indústria em tempo de guerra*, 36.

71. See Ferreira, "Nacionalização e confisco do capital português," on the nationalization of the industrial sector.

72. João Filipe Martins, Ministério da Informação, divulagação do despacho no. 1, 28 November 1975, RNA arquívo histórico.

73. Lopes, "A evolução da radiodifusão em Angola," 45. Interviews by author in Luanda: Guilherme Mogas, 11 and 24 May 2011; Elisabete Pereira, 11 April 2011; Reginaldo Silva, 24 January 2011.

74. Ladislau Silva spent six weeks in Czechoslovakia in the early 1980s (interview by author, 16 February 2011, Luanda); Reginaldo Silva spent three months in East Germany (interview by author, 24 January 2011, Luanda); Carlos Monteiro Ferreira spent four months in East Germany (interview by author, 16 March 2011, Luanda); José Patrício spent a year in Belgrade, Yugoslavia, at the Instituto Superior of Journalism (interview by author, 13 June 2012, Luanda); Filipe Lomboleni completed an advancement course in Czechoslovakia; in 1984 he went with the Novost Soviet international news agency to Moscow and Belorus (interview by author, 30 May 2011, Luanda); Sebastião Silva completed a three-month course on sound in East Germany (interview by author, 20 April 2011, Luanda); Elisabete Pereira spent fifteen days in Cuba in 1983, after which she developed child programming (interveiw by author, 11 April 2011, Luanda).

75. RNA, Departamento da Informação, Relatório Mensal, September 1980, regarding reporter and editor Bernardo Neto's departure for Cuba for a journalism course. Documents from the collection of Elisabete Pereira contain Cuban-produced instructional material on the basic practice of journalism and news writing by the Cuban Union of Journalists. "Por uma imprensa ao serviço da revolução II: Dados básicos para a prática do jornalismo" (Edição Angop/DOR, n.d.), 8 pp. (from Ricardo Cardet, "Lição de prática de jornalismo," in Teoria de Técnica, Material Docente de União de Jornalistas de Cuba, n.d.); and Benitez, "Por uma imprensa ao serviço da revolução I: Origens, significado e valor de uma notícia" (Edição Angop/DOR, n.d.), 31 pp. (from "La noticia," in Tecnica Periodistica, Union de Periodistas de Cuba, 1971). Both from Elisabete Pereira, personal collection.

76. "Locução de noticiários," 3pp.; and "Curso de superação dos locutores daRNA," 28 pp., n.d., Elisabete Pereira, personal collection.

77. The Spanish material, titled "Radio Information," described radio as the guerrilla troops to the heavy artillery, infantry, and air force, of other forms of social communication (press, television, weekly magazines). A document from RFI, "Fazer rádio?

Sim, mas para quê?" (Radio? Yes, but why?), looks at the who, what, why, and how of programming and production. Elisabete Pereira, personal collection.

78. "Leadership," Elisabete Pereira, personal collection.

79. Departamento da Informação, Relatório Mensal, June 1980, RNA. Well aware of the security issues related to the president's movements in a time of civil war, they point out that the country's first president, Agostinho Neto, always gave them ample time to prepare.

80. Departamento da Informação, Relatório Mensal, May 1980, RNA.

81. Departamento da Informação, Relatório Mensal, May 1980, RNA.

82. Radialist António Fonseca remembers working at Rádio Ecclesia, the Catholic station, that day. When he heard *Kudibangela* on Ecclesia that morning (it was broadcasting in network with RNA as usual), he knew "something was wrong." He took the risk of cutting the link with RNA and notifying the party. Ecclesia went silent. And then broadcast only music. Listeners knew that whatever was happening at RNA, they did not control the airwaves completely. At 1 p.m., Ecclesia broadcast the official announcement of the attempted coup, phoned in by party leader Lúcio Lara and read by António Fonseca. As part of the centralization that followed the 27 de Maio, the MPLA closed Ecclesia. António Fonseca, interview by author, 23 April 2011, Luanda.

83. António Fonseca, interview by author, 23 April 2011, Luanda. See, too, Coelho, *Angola*, 126, 173–76. According to Sebastião Coelho, he is responsible for the Angolanization of radio with the inauguration of his program *Cruzeiro do Sul*, directed to African audiences and broadcast from Rádio Clube de Huambo in Umbundo. After a stint in jail thanks to the PIDE, he started broadcasting *Tondoya Mukina o Kizomba*, a music show based in Luanda via his production company Estúdios Norte. Coelho, *Angola*, 194–98.

84. António Fonseca, interview by author, 23 April 2011, Luanda. Fonseca points to the new sound and programming instituted by the youth of the period, trained at Rádio Escola, including Silva Júnior, João das Chagas, Carlos Ferreira, and Alves Fernandes. Other, older employees did not see the importance of African cultural practices and history. Nonetheless, RNA continues to use continental and not Angolan Portuguese for official radio speech.

85. The program *Boa Noite Angola*, likely from 1984, announced new hours, three hours, from 10 p.m. until 1 a.m. From the RNA, Historical Archive, 27 May 2011, 27m47s.

86. Ladislau Silva, interview by author, 11 February 2011, Luanda.

87. Debra Spitulnik has written about this service at the Zambian National Broadcasting Corporation. Spitulnik, "Personal News and the Price of Public Service: An Ethnographic Window into the Dynamics of Production and Reception in Zambian State Radio," in *The Anthropology of News and Journalism: Global Perspectives*, ed. S. Elizabeth Bird (Bloomington: Indiana University Press, 2009), 182–98.

88. Ladislau Silva, interview by author, 16 February 2011, Luanda.

89. Ferreira, *A indústria em tempo de guerra*, 67–68.

90. Interviews by author in Luanda: Reginaldo Silva, 24 January 2011; and Ladislau Silva, 16 February 2011.

91. Ladislau Silva, interview by author, 16 February 2011, Luanda.

92. For the history of music production in Angola, see Coelho, *Angola*; and Moorman, *Intonations*. Coelho's record company, Companhia de Discos de Angola

(CDA; Angolan Record Company), was taken over by the state's Instituto Nacional de Angola de Livros e Discos (INALD; Angolan National Institute of Books and Records). See Coelho, *Angola*, 209–16, for an explanation of the tangled process.

93. Interviews by author in Luanda: Carlos Monteiro Ferreira, 16 March 2011; Bruno Lara, 17 April 2011; and Eduardo Paim, 9 May 2011.

94. Bruno Lara, interview by author, 17 April 2011, Luanda.

95. RNA documents of Elisabete Pereira, personal collection.

96. Interviews by author in Luanda: Guilherme Mogas, 11 May 2011; and Carlos Monteiro Ferreira, 16 March 2011.

97. Guilherme Mogas, interview by author, 11 May 2011, Luanda.

98. José Eduardo dos Santos trained as a petroleum engineer in Baku, Azerbaijan, then the USSR, between 1963 and 1969. See http://www.embaixadadeangola.org/politica02.htm.

99. Reginaldo Silva, interview by author, 24 January 2011, Luanda. Silva started work at the radio in the transitional period when it was still EOA in September or October 1975. In the middle of 1976, he left to work at the newspaper *Diário de Luanda*, which was later closed by state dispatch for its association with Nitistas. Silva went to *Jornal de Angola* in early 1977. He was arrested on 28 May 1977 and jailed for two years. Rui de Carvalho requested that the MPLA readmit him to employment at RNA, and he returned in August 1979, after having been told he could never again work in the area of communications and being sent to work in the Ministry of Transport in Angola's easternmost Lunda Norte Province.

100. Reginaldo Silva, interview by author, 24 January 2011, Luanda.

101. Ladislau (Lau) Silva, 16 February 2011, Luanda.

102. Guilherme Mogas, interview by author, 11 May 2011, Luanda.

103. Sebastião Coelho is unsparing in his criticism of the effects of RNA's monopoly and centralization. He says that RNA prioritizes Luanda and its news and happenings. Coelho, *Angola*, 121–22. Filipe Lombelini, director of the Department of National Languages, made a similar point more elliptically when he said that Angola needed community radio stations throughout the country for radio to truly be national. Filipe Lomboleni, interview by author, 30 May 2011, Luanda.

104. Guilherme Mogas, interview by author, 11 May 2011, Luanda.

105. Guilherme Mogas, interview by author, 11 May 2011, Luanda. See Ferreira, *A indústria em tempo de guerra*, 67–68, on the question of imports and the economic emergency of the early 1980s.

106. Interviews by author in Luanda: Reginaldo Silva, 24 January 2011; and Carlos Monteiro Ferreira, 16 March 2011. Tali notes that the MPLA party opened the Escola do Partido (Party School) after the 1° Congresso in 1977 to provide political education in Marxist-Leninist theory. Tali, *Guerrilhas e lutas sociais*, 615.

107. Guilherme Mogas, interview by author, 11 May 2011, Luanda.

108. Guilherme Mogas, interview by author, 11 May 2011, Luanda.

109. Guilherme Mogas, interview by author, 11 May 2011, Luanda.

110. See Ferreira, "Nacionalização e confisco do capital português."

111. Interviews by author in Luanda: Guilherme Mogas, 11 and 24 May 2011; and Carlos Monteiro Ferreira, 16 March 2011.

112. Matthew Connelly analyzes the FLN's brilliant use of such a strategy in *A Diplomatic Revolution: Algeria's Fight for Independence and the Origins of the Post–Cold War Era* (Oxford: Oxford University Press, 2003).

CHAPTER 5: "ANGOLA: THE FIRM TRENCH OF THE REVOLUTION IN AFRICA!"

1. See https://www.youtube.com/watch?v=7ydoo6OdMy4.

2. Lennart Bolliger, "Apartheid's Transnational Soldiers: The Case of Black Namibian Soldiers in South Africa's Former Security Forces," *Journal of Southern African Studies* 43, no. 1 (2017): 195–214; Bolliger, "Chiefs, Terror, and Propaganda: The Motivations of Namibian Loyalists to Fight in South African Security Force, 1975–1989," *South African Historical Journal* 70, no. 1 (2018): 124–51; and Bolliger "The Origins and Formation of 32 Battalion, 1975–76" (chap. 3 of unpublished thesis, cited with author's permission); Erik Kennes and Miles Larmer, *The Katangese Gendarmes and War in Central Africa: Fighting Their Way Home* (Bloomington: Indiana University Press, 2016); Bernice Labuschagne, "South Africa's Intervention in Angola: Before Cuito Cuanavale and Thereafter" (master's thesis, Stellenbosch University, 2009); John Stockwell, *In Search of Enemies: A CIA Story* (New York: Norton, 1978).

3. On the oil boom, see Ricardo Soares de Oliveira, *Magnificent and Beggar Land: Angola since the Civil War* (London: Hurst, 2015); and on the politico-cultural system, see Jon Schubert, *Working the System: A Political Ethnography of the New Angola* (Ithaca, NY: Cornell University Press, 2017).

4. Peter Baxter, *SAAF's Border War: The South African Air Force in Combat, 1966–1989* (Pinetown: 30 Degrees South Publishers, 2012); Jannie Geldenhuys, *A General's Story: From an Era of War and Peace* (Johannesburg: Jonathan Ball, 1995); Piero Gleijeses, *Visions of Freedom: Havana, Washington, Pretoria, and the Struggle for Southern Africa, 1976–1991* (Chapel Hill: University of North Carolina Press, 2013); Labuschagne, "South Africa's Intervention in Angola"; and Karl Maier, *Angola: Promises and Lies* (Rivonia: William Waterman, 1996). W. Martin James III, *A Political History of the Civil War in Angola, 1974–1990* (New Brunswick, NJ: Transaction Publishers, 2011); Pombo [Harry Villegas], *Cuba and Angola: The War for Freedom* (New York: Pathfinder, 2017), 69–76.

5. Sue Onslow, ed., *Cold War in Southern Africa: White Power, Black Liberation* (New York: Routledge, 2009).

6. Onslow, *Cold War in Southern Africa*, 6.

7. Jeffrey Herf, "Narrative and Mendacity: Anti-Semitic Propaganda in Nazi Germany," in *The Oxford Handbook of Propaganda Studies*, ed. Jonathan Auerbach and Russ Castronovo (New York: Oxford University Press, 2013), 91.

8. Herf, "Narrative and Mendacity," 92.

9. On how radio structured the rhythms of the day, marking differences between work and leisure, see Brian Larkin, *Signal and Noise: Media, Infrastructure, and Urban Culture in Nigeria* (Durham, NC: Duke University Press, 2008), 52.

10. Jo Tacchi, "Radio and Affective Rhythm in the Everyday," *Radio Journal* 7, no. 2 (2009): 171–83.

11. "Presidente dos Santos falando das trop Cubanos e indep Namibia," RNA Arquivo Histórico, n.d.

12. "Presidente dos Santos falando das trop Cubanos e indep Namibia," RNA Arquivo Histórico, n.d.

13. See, for example, Matthew Connelly, *A Diplomatic Revolution: Algeria's Fight for Independence and the Origins of the Post–Cold War Era* (Oxford: Oxford University Press, 2003); Gabrielle Hecht, ed., *Entangled Geographies: Empire and Technopolitics*

in the Cold War (Cambridge, MA: MIT Press, 2011); Vijay Prashad, *The Darker Nations: A People's History of the Third World* (New York: New Press, 2008); and Odd Arne Westad, *The Global Cold War: Third World Interventions and the Making of Our Times* (New York: Cambridge University Press, 2005).

14. Christopher J. Lee, ed., *Making a World after Empire: The Bandung Moment and Its Political Afterlives* (Athens: Ohio University Press, 2010), 19.

15. Christopher J. Lee, "Decolonization of a Special Type: Rethinking Cold War History in Southern Africa," *Kronos* 37, no. 1 (2011): 9.

16. Maria Paula Meneses and Bruno Sena Martins, *As guerras de libertação e sonhos coloniais: Alianças secretas, mapas imaginados* (Lisbon: Almedina, 2013), 17. Lee points to the difficulty these relations pose to historians of southern Africa and how they do not fit the Manichaean divisions of colonizer/colonized. Lee, "Decolonization of a Special Type," 8.

17. See, for example, the following chapters in Meneses and Martins, *As guerras de libertação e sonhos coloniais*: Maria Paula Meneses and Catarina Gomes, "Regressos? Os *retornados* na (des) colonização portuguesa," 59–108; Carlos de Matos Gomes, "A africanização na guerra colonial e as suas sequelas tropas locais: Os vilões nos ventos da história," 123–41; and Maria Paula Meneses, Celso Braga Rosa, and Bruno Sena Martins, "Estilhaços do *Exercício Alcora*: O epílogo dos sonhos coloniais," 171–79.

18. Gary Baines, *South Africa's "Border War": Contested Narratives and Conflicting Memories* (London: Bloomsbury, 2014); Gary Baines and Peter Vale, eds., *Beyond the Border War: New Perspectives on Southern Africa's Late Cold War Conflicts* (South Africa: UNISA Press, 2008); Piero Gleijeses, *Conflicting Missions: Havana, Washington, and Africa, 1959–1976* (Chapel Hill: University of North Carolina Press, 2002); and Gleijeses, *Visions of Freedom*; Jose Milhazes, *Angola: O princípio do fim da União Sovietica* (Lisbon: Vega, 2009); Onslow, *Cold War in Southern Africa*; Vladimir G. Shubin, *The Hot "Cold War": The USSR in Southern Africa* (London: Pluto Press, 2008); Westad, *Global Cold War.*

19. Onslow, *Cold War in Southern Africa*, 25.

20. Christian A. Williams, "Living in Exile: Daily Life and International Relations at SWAPO's Kongwa Camp," *Kronos* 37, no. 1 (2011): 60–86.

21. Onslow, "The Cold War in Southern Africa: White Power, Black Nationalism and External Intervention," in Onslow, *Cold War in Southern Africa*, 19–20.

22. The PIDE's murders of Amílcar Cabral and Eduardo Mondlane, with collusion of PAIGC and Frelimo members, loomed large. See Irene Flunser Pimentel, *A história da PIDE* (Lisbon: Circulo de Leitores, 2011), 411; Dalila Cabrita Mateus, *A PIDE/DGS na guerra colonial, 1961–1974* (Lisbon: Terramar, 2004), 165–72; and António Tomás, *O fazedor de utopias: Uma biografia de Amílcar Cabral* (Lisbon: Tinta da China, 2007), 265–86.

23. Clapperton Mahvunga, "A Plundering Tiger with Its Deadly Cubs? The USSR and China as Weapons in the Engineering of a 'Zimbabwean Nation,' 1945–2009," in Hecht, *Entangled Geographies*, 231–66.

24. Interviews by author in Luanda: Paulo Jorge, 2 August 2005; Guilherme Mogas, 24 May 2011; Reginaldo Silva, 24 January 2011; Carlos Monteiro Ferreira, 16 March 2011.

25. Interviews by author in Luanda: Bruno Lara, 17 April 2011; and Guilherme Mogas, 11 and 24 May 2011.

26. Mahvunga, "Plundering Tiger."

27. Paulo de Carvalho and Reginaldo Silva, *Estado da comunicação social em Angola* (Luanda: Conselho Nacional de Comunicação Social, 2007), 14–15.

28. For some examples, see Sousa Jamba, *Patriots* (London: Viking, 1990); Ondjaki, *Good Morning Comrades: A Novel*, trans. Stephen Henighan (Emeryville: Biblioasis, 2001), originally published as *Bom dia camaradas: Romance* (Luanda: Chá de Caxinde, 2000); and Ondjaki, *Os Transparentes* (Lisbon: Caminho, 2012); Manuel Rui, *Quem me dera ser onda* (Lisbon: Cotovia, 1991).

29. MPLA-Partido do Trabalho, *Relatório do Comite Central ao II Congresso do MPLA-Partido do Trabalho* (Luanda: Makutanga, 1986), 38, AO-Pol IV-2 dossier, Centro de Informação Amílcar Cabral, Lisboa. On audience, Reginaldo Silva, conversation with author, 19 July 2018, Luanda.

30. Reginaldo Silva, conversation with author, 19 July 2018, Luanda.

31. MPLA-PT, "Teses e resoluções, 1° Congresso" (Luanda: Imprensa Nacional, 1978), 109.

32. MPLA-PT, "Teses e resoluções, 1° Congresso" (Luanda: Imprensa Nacional, 1978), 114.

33. Paulo Jorge, interview by author, 2 August 2005, Luanda.

34. Paulo Jorge, interview by author, 2 August 2005, Luanda.

35. Matthew Connelly, "Rethinking the Cold War and Decolonization: The Grand Strategy of the Algerian War for Independence," *International Journal of Middle East Studies* 33, no. 2 (2001): 223; and Connelly, *Diplomatic Revolution*, 24.

36. Delinda Collier, "A 'New Man' for Africa? Some Particularities of the Marxist *Homem Novo* within Angolan Cultural Policy," in *De-Centering the Cold War History: Local and Global Change*, ed. Jadwiga E. Pieper Mooney, and Fabio Lanza (New York: Routledge, 2013), 190.

37. Collier, "A 'New Man' for Africa," 196–98.

38. MPLA-PT, "Teses e resoluções, 1° Congresso," 101–25.

39. MPLA-PT, "Teses e resoluções, 1° Congresso," 109.

40. *Jornal de Angola*, "MPLA (Documentos e Teses ao 1° Congresso)," December 1977, n.p.

41. MPLA-PT, "Teses e resoluções, 1° Congresso," 115.

42. MPLA-PT, "Teses e resoluções 1° Congresso," 115.

43. This included Ruth Lara, wife of Neto's secondhand man, Lúcio Lara, and Costa Andrade "Ndunduma," a writer and the editor of *Jornal de Angola*. Ndunduma wrote this play about a painting of the president and a group performed it on the president's birthday. Meant to be a "gift," the president did not find it amusing.

44. Mbeto Traça advanced to become a military general in the Angolan Armed Forces and Paulo Jorge served for many years as the foreign relations minister. Adolfo Maria was marginalized and persecuted for his role in the Revolta Activa. He went into hiding for many years and then exile. See Fernando Tavares Pimenta, *Angola no percurso de um nacionalista: Conversas com Adolfo Maria* (Lisbon: Edições Afrontamento, 2006); and Marcelo Bittencourt, *"Estamos juntos!": O MPLA e a luta antiocolonial, 1961–1974*, vol. 2 (Luanda: Kilombelombe, 2008"), 225–39; and Jean-Michel Mabeko Tali, *O MPLA perante si próprio*, vol. 2 of *Dissidências e o poder de estado (1962–1977)* (Luanda: Nzila Editora, 2001), 184–90.

45. FBIS-SSA-76-106, 6 January1976, Luanda Domestic Service in Portuguese.

46. Wilfred G. Burchett and Derek Roebuck, *The Whores of War: Mercenaries Today* (London: Penguin, 1977).

47. They appear as background characters in José Eduardo Agualusa's *Estação das Chuvas* (Lisbon: Dom Quixote, 2007).

48. *O Povo Acusa* (Luanda, 1976), penultimate page, pages not numbered.

49. "Death for 'War Dogs,'" *Time* magazine, 12 July 1976, 42.

50. *O Povo Acusa* (Luanda, 1976), penultimate page, pages not numbered.

51. Burchett and Roebuck, *Whores of War.*

52. Arquivo de Associação Tchiweka, Mercenaries subfile: minutes of antimercenary group meeting, without date but likely 1976, 2pp.; antimercenaries liaison group meeting, 2 pp., 29 October 1976.

53. Gary Baines, "The Saga of South African POWs in Angola, 1975–82," *Scientia Militaria* 40, no. 2 (2012): 105. Stockwell, *In Search of Enemies*, 185–87.

54. FBIS-SAF-75-244, 18 December 1975, Johannesburg International service in English.

55. Baines, "The Saga of South African POWs in Angola," 107–8. The Nigerian president Muhammed Murtala had decided to support the struggle in southern Africa as of July 1975. Author's personal communication with Akin Adesokan.

56. Gleijeses, *Conflicting Missions*, 323–27; Gleijeses notes that even the African presses, like the Nigerian and the Tanzanian, both hostile to South Africa, hesitated to report the news of South African troops on the ground. The lack of foreign observers and journalists on the ground made reporting extremely difficult, as did pressure on US news sources not to report it.

57. For Onslow, 1976 was a watershed year in the region for this reason. Onslow, *Cold War in Southern Africa*, 16.

58. On South African media and propaganda in this period, see Baines, "Saga of South African POWs in Angola"; on speaking to a white electorate, see Keyan G. Tomaselli and Ruth Tomaselli, *Media Reflections of Ideology* (Durban: University of Natal, 1985).

59. Ondjaki, *Good Morning Comrades*, 19. For the same practice in the colonial period, see Ana Sofia Fonseca, *Angola, terra prometida: A vida que os portugueses deixaram* (Lisbon: Esfera dos Livros, 2011), 251, where she says: "Lunchtime is the noble hour. Days run by serenely, at mealtimes all reunite at home, around the table. The radio is part of the family." In José Eduardo Agualusa's *A teoria geral do esquecimento* (Lisbon: Dom Quixote, 2012), 57, the mercenary Jeremias is hidden for five months and depends on the radio to follow developments of the war between government troops, with Cuban help, against UNITA, the FNLA, South African troops, Portuguese, English, and American mercenaries.

60. Tacchi, "Radio and Affective Rhythm in the Everyday." Tacchi discusses the way that individuals use radio to manage moods and how radio helps structure domestic soundscapes.

61. Miguel Júnior, *A mão Sul-Africana O envolvimento das forças de defesa da África do Sul no sudeste de Angola (1966–1974)* (Caiscais: Tribuna de História, 2014), 61–66.

62. Oscar Cardoso, interview by author, 25 January 2016, Ericeira; and William Minter, *Apartheid's Contras: An Inquiry into the Roots of War in Angola and Mozambique* (Atlantic Highlands, NJ: Zed Books, 1994), 231.

63. See Bolliger's forthcoming dissertation chapter on South Africa's Buffalo Battalion, "The Origins and Formation of 32 Battalion, 1976–76."

64. For work on the ANC in exile and particularly in Angola, a contentious, troubled history, see Todd Cleveland, "'We Still Want the Truth': The ANC's Angolan Detention Camps and Post-Apartheid Memory," *Comparative Studies of South Asia, Africa, and the Middle East* 25, no. 1 (2005): 63–78; Stephen Ellis, "The ANC in

Exile," *African Affairs* 90, no. 360 (1991): 439–47; and Paul Trewhela, *Inside Quatro: Uncovering the Exile History of the ANC and SWAPO* (Auckland Park: Jacana, 2009).

65. Piero Gleijeses, *Visions of Freedom*, 10.

66. Gleijeses, *Conflicting Missions*, 275–76, 295; and Minter, *Apartheid's Contras*, 188–91.

67. United Nations Center against Apartheid, International Commission of Inquiry, *Acts of Aggression Perpetrated by South Africa against the People's Republic of Angola, June 1979–July 1980* (New York: United Nations, 1981), 2.

68. UN Security Council Resolutions S/387 (1976), S/428 (1978), S/447 (1979), S/454 (1979), and S/475 (1980), available at http://www.un.org/en/sc/documents/resolutions/.

69. UN Security Council Resolution S/435 (1978), http://www.un.org/en/sc/documents/resolutions/.

70. United Nations Center against Apartheid, *Acts of Aggression*, 6.

71. United Nations Center against Apartheid, *Acts of Aggression*, 8.

72. MPLA-Partido do Trabalho, *Teses e Resoluções 1º Congresso* (Luanda: Imprensa Nacional de Angola, 1978), 108.

73. MPLA-Partido do Trabalho, *Teses e Resoluções 1º Congresso*, 111.

74. RNA, News Flashes, 23h00, 6 August 1980.

75. RNA, News Flashes, 15 July 1980.

76. RNA, News Flashes, 21 May 1981.

77. RNA, News Flashes, 27 August 1981.

78. RNA, News Flashes, 6 September 1981.

79. For example, see RNA, News Flashes, March 1982.

80. RNA, News Flashes, 13 April 1980, 13h00.

81. RNA, News Flashes, 12 March 1982.

82. RNA, News Flashes, 3 September 1982.

83. RNA, News Flashes, 5 August 1980.

84. RNA News Flash, 9 July 1980.

85. RNA News Flash, 3 August 1980.

86. RNA News Flash, 15 August 1982.

87. RNA News Flash, 17 June 1980.

88. RNA News Flash, November 1980.

89. Baxter, *SAAF's Border War*, 52–58; Gleijeses, *Visions of Freedom*, 424; Shubin, *Hot "Cold War,"* 112.

90. Baxter, *SAAF's Border War*, 50–61; Gleijeses, *Visions of Freedom*, 393–430. Shubin acknowledges it but plays it down in *Hot "Cold War,"* 107.

91. Gleijeses, *Visions of Freedom*, 425–26; and James, *Political History of the Civil War in Angola*, 177, 218–19.

92. James, *Political History of the Civil War in Angola*, 425–26.

93. Baxter, *SAAF's Border War*, 59.

94. CNN interview with José Eduardo dos Santos, "Pres dos Santos entrevista a indep da Namibia 1988," n.d., Arquivo Histórico, Rádio Nacional de Angola.

95. See Aharon De Grassi for a more complete discussion of this process and its relations to the IMF. De Grassi, "Provisional Reconstructions: Geo-Histories of Infrastructure and Agrarian Configuration in Malanje, Angola" (PhD diss., University of California, Berkeley, 2017), 84.

96. RNA documents, DIOP (Departamento de Intercâmbio e Opinião Pública/ Department of Exchange and Public Opinion), "Inquerito 1986," June 1986, 43 pp.

The sample size was small (700 people) and came only from Luanda and neighboring Bengo provinces. They expressed hope to expand the survey later in 1986, but I found no reports from then.

97. See the work of Fernando Alvim, *Memórias, marcas, íntimas*, mixed-media installation. 1996; Victor Gama, "Tectonik Tombwa" and "Vela 69," http://www.victorgama.org/projects.html; Cedric Nunn, https://www.cedricnunn.co.za/essays/cuito-cuanavale.html; and Jo Ractliffe, *As terras do fim do mundo*, http://archive.stevenson.info/exhibitions/ractliffe/index2010.htm.

CHAPTER 6: RADIO VORGAN

1. Sousa Jamba, *Patriots* (London: Viking, 1990), 286.

2. Franco Marcolino, Radio Vorgan, personal document; and Nick Grace C., "Investigative Report: Vorgan; A Voz de Resistência do Galo Negro—A Radio of Africa's Longest-Running War," 27 February 1998; updated 4 April 1998, p. 1, on Clandestine Radio Intel website, http://www.qsl.net/ybormi/vorgan.htm.

3. Elaine Windrich, "The Laboratory of Hate: The Role of Clandestine Radio in the Angolan War," *International Journal of Cultural Studies* 3, no. 2 (2000): 206–18. She takes this description from Anita Coulson's sidebar on UNITA in "Angola: Constructing Capitalism," *Africa Report* 36, no. 6 (November 1991): 63. What Coulson attributes to the government, Windrich generalizes to "known in Luanda" (210).

4. John A. Marcum, *The Anatomy of an Explosion (1950–1962)*, vol. 1 of *The Angolan Revolution*; and Christine Messiant, *L'Angola colonial, histoire et société: Les prémisses du mouvement nationaliste* (Switzerland: P. Schlettwein, 2006), 98–104.

5. Maria da Conceição Neto's "In and Out of Town"; and Didier Péclard, *Les incertitudes de la nation en Angola: Aux racine sociales de l'Unita* (Paris: Karthala, 2015).

6. Péclard, *Les incertitudes de la nation en Angola*, 133–34.

7. Key texts include: Jean Comaroff and John L. Comaroff, *Of Revelation and Revolution*, vol. 1, *Christianity, Colonialism, and Consciousness in South Africa* (Chicago: University of Chicago Press, 1991); and Comaroff and Comaroff, *Of Revelation and Revolution*, vol. 2, *The Dialectics of Modernity on a South African Frontier* (Chicago: University of Chicago Press, 1997); Paul Landau, *The Realm of the Word: Language, Gender and Christianity in a Southern African Kingdom* (Portsmouth, NH: Heinemann, 1995); and Derek R. Peterson, *Creative Writing: Translation, Bookkeeping, and the Work of Imagination in Colonial Kenya* (Portsmouth, NH: Heinemann, 2004); and Peterson, *Ethnic Patriotism and the East African Revival: A History of Dissent, c. 1935–1972* (Cambridge: Cambridge University Press, 2014).

8. See Péclard, *Les incertitudes de la nation en Angola*, chap. 4.

9. Péclard, *Les incertitudes de la nation en Angola*, 239.

10. Maria da Conceição Neto, "In Town and Out of Town: A Social History of Huambo (Angola), 1902–1961" (PhD thesis, SOAS, University of London, 2012), 218.

11. Neto mentions the Catholic seminary at Kwima near Huambo. Neto, "In Town and Out of Town," 217.

12. Neto, "In Town and Out of Town," 208, 217.

13. Neto points out that education at a Protestant mission resulted in recognition for only the first two years of secondary school, and the second two had to be completed at a government institution. Neto, "In Town and Out of Town," 220.

14. Neto, "In Town and Out of Town," 224.

15. Jean-François Bayart and Stephen Ellis, "Africa in the World: A History of Extraversion," *African Affairs* 99, no. 395 (2000): 217–67.

16. Péclard, *Les incertitudes de la nation en Angola*, chap. 3 and 4, 239. Linda M. Heywood and John K. Thornton's work formulates this in terms of an Atlantic Creole culture. See Heywood and Thornton, *Central Africans, Atlantic Creoles, and the Foundation of the Americas, 1585–1660* (New York: Cambridge University Press, 2007); and John K. Thornton, *A Cultural History of the Atlantic World, 1250–1820* (New York: Cambridge University Press, 2012).

17. Clapperton Mavhunga, "Usage Engineers as a Force behind the Movement of Technology, Part 1: *Guns in Pre-Colonial Africa* (presentation at Department of Science and Technology Studies Seminar Series, Cornell University, September 2012), cited with author's permission.

18. Neto, "In Town and Out of Town," 238.

19. Jamba, *Patriots*, 17, 25, 109, 116, 164, 190, 217, 253.

20. Jamba, *Patriots*, 12, 80, 81, 85, 164, 198, 234, 251, 255, 282, 286, 288.

21. On representations of Savimbi in the US, see Piero Gleijeses, *Visions of Freedom: Havana, Washington, Pretoria, and the Struggle for Southern Africa, 1976–1991* (Chapel Hill: University of North Carolina Press, 2013), 296–99; Tony Hodges, *Angola: Anatomy of an Oil State* (first published as *Angola: From Afro-Stalinism to Petro-Diamond Capitalism*) (Bloomington: Indiana University Press, 2001), 18, 47; Bela Malaquias, interview by author, 6 June 2008, Luanda.

22. Bela Malaquias, interview by author, 6 June 2008, Luanda.

23. See Fred Bridgland on Tito Chingunji and the story of how Savimbi first used a young woman, Ana, to seduce Chingunji and then took her as his own concubine to challenge Chingunji psychologically. Bridgland, "Death in Africa," Cold Type, 2004, coldtype.net.

24. Samuel Chiwale, *Cruzei-me com a história* (Lisbon: Sextante Editora, 2008), 259–75.

25. Lourenço Bento, interview by author, 28 January 2011, Luanda. One of Pearce's informants recalled cassette recordings of Savimbi's speeches being brought to Huambo in the 1980s and UNITA supporters listening to them in secret. Justin Pearce, personal communication with author, May 2016.

26. KUP, 2 November 1988, FBIS-AFR, 88–212.

27. Among UNITA leadership, Savimbi's murder of Tito Chingunji marked a turning point, particularly relative to international support. Savimbi's former biographer Fred Bridgland finally spoke out about Chingunji's murder and former supporters in Washington, DC, where Chingunji was much respected and beloved, experienced some disillusion, even if it did not occasion an immediate change in policy in the region. See Bridgland, "Death in Africa," Cold Type, 2004, 1–50, coldtype.net; Greg Pirio, interview by author, 18 August 2015, via Skype.

28. Gleijeses, *Visions of Freedom*, 192.

29. Windrich, "Laboratory of Hate," 208.

30. At this point Vorgan consisted of Olinda Colanda, Bela Malaquias, Franco Marcolino, Vasco Sassango, Xico Torres, Wilson dos Santos, and Zito Calhas. Interviews by author: Bela Malaquias, 6 June 2008, Luanda; and Miraldina Jamba, 18 February 2011, Luanda.

31. Windrich, "Laboratory of Hate," 208.

32. Windrich, "Laboratory of Hate," 207–9.

33. Lourenço Bento, interview by author, 28 January 2011, Luanda.

34. Lourenço Bento, interview by author, 28 January 2011, Luanda.

35. Bela Malaquias interview by author, 6 June 2008, Luanda. Malaquias dates the move to 1981.

36. Interviews by author: Bela Malaquias 6 June 2008, Luanda; Lourenço Bento, 28 January 2011, Luanda; and Armando Ferramenta, 8 February 2011, Luanda. In 1988 Greg Pirio, then director of the Voice of America's Portuguese broadcasting to Africa, visited Luanda, and RNA director Mogas asked him why Vorgan broadcast on the VOA frequency. Pirio, unaware that this was the case, but clear that this was illegal according to VOA's charter, had no answer. In an interview later that year with Savimbi in Washington, DC, Pirio was told that UNITA handled only the programming, and the CIA set up the broadcasting. Interviews by author: Guilherme Mogas, 24 May 2011, Luanda; and Gregory Pirio, 18 August 2015, via Skype.

37. Raúl Manuel Danda, interview by author, 10 February 2011, Luanda.

38. Lourenço Bento, interview by author, 28 January 2011, Luanda.

39. Lourenço Bento, interview by author, 28 January 2011, Luanda.

40. Raúl Manuel Danda, interview by author, 10 February 2011, Luanda.

41. Bela Malaquias, interview by author, 6 June 2008, Luanda.

42. Miraldina Jamba, interview by author, 18 February 2011, Luanda.

43. FBIS-AFR-88-113, KUP, 13 June 1988.

44. Paulo de Carvalho, *Audiência de media em Luanda* (Luanda: Coleção Ensaio, 2002), 103. In casual conversation, people mentioned listening to UNITA during the war years.

45. Lourenço Bento, interview by author, 28 January 2011, Luanda.

46. Wilson dos Santos, Almerindo Jaka Jamba, and Jorge Valentim served as secretaries of information; and Franco Marcolino "Nhany" (also one of its best-known announcers) in the early days, Moises Armindo Kasesa, João Evangelista (who had experience from Rádio Clube de Saurimo), Zito Calhas, and Clarindo Kaputo all served as directors at one time. Interviews by author: Lourenço Bento, 28 January 2011, Luanda; Armando Ferramenta, 8 February 2011, Luanda; and Miraldina Jamba, 18 February 2011, Luanda.

47. Savimbi sponsored these educational seminars organized by the information secretaries Almerindo Jaka Jamba and then Jorge Valentim, and by the Vorgan directors Clarindo Kaputo and later Franco Marcolino. Interviews by author: Lourenço Bento, 28 January 2011, Luanda; Armando Ferramenta, 8 February 2011, Luanda; and Ricardo Branco, personal communication with author, 31 May 2013.

48. Ricardo Branco, personal communication with author, 31 May 2013.

49. Ricardo Branco, personal communication with author, 31 May 2013.

50. For work on the Portuguese who left Angola and Mozambique for South Africa, see Pamila Gupta, "'Going for a Sunday Drive': Angolan Decolonization, Learning Whiteness and the Portuguese Diaspora of South Africa," in *Narrating the Portuguese Diaspora: Piecing Things Together*, ed. Francisco Cota Fagundes, Irene Maria F. Blayer, Teresa F. A. Alves, and Teresa Cid (New York: Peter Lang, 2011), 135–52; and Clive Glaser, "The Making of a Portuguese Community in South Africa (1900–1994)," in *Imperial Migrations: Colonial Communities and Diaspora in the Portuguese World*, ed. Eric Morier-Genoud and Michael Cahen (New York: Palgrave Macmillan, 2012), 223–24. In Portugal most of the work on *retornados* is in fiction, but see Maria Paula Meneses and Catarina Gomes, "Regressos? Os *retornados* na (des) colonização

portuguesa," in *As guerras de libertação e os sonhos coloniais: Alianças secretas, mapas imaginados* (Lisbon: Almedina, 2013), 59–108.

51. Oscar Cardoso, interview by author, 25 January 2016, Ericeira. Cardoso worked with Ken Flowers in Rhodesia.

52. Bela Malaquias, interview by author, 6 June 2008, Luanda.

53. Lourenço Bento, interview by author, 28 January 2011, Luanda.

54. Radio Vorgan, document written by and sent to me by Franco Marcolino.

55. Pepetela, *A Geração da Utopia* (Lisbon: Dom Quixote, 1997).

56. Lourenço Bento, interview by author, 28 January 2011, Luanda. The start of this show was announced on Vorgan in late March 1983: "The Voice of the Resistance of the Black Cockerel will soon introduce a program specifically dedicated to FALA fighters. This program will have a formative and informative mission and—why not?—include our good music, which will be carefully selected so that it is mixed with words and so that the combination provides happiness, color, and music and, above all, enables FALA fighters to reflect on life." FBIS-MEA-83-059, Voice of Resistance of the Black Cockerel, 25 March 1983.

57. Lourenço Bento, interview by author, 28 January 2011, Luanda.

58. Lourenço Bento, interview by author, 28 January 2011, Luanda.

59. Raúl Manuel Danda, interview by author, 10 February 2011, Luanda.

60. KUP's founding was announced on Vorgan 10 October 1984. FBIS-MEA-84-197, Voice of Resistance of the Black Cockerel, 10 October 1984.

61. Interviews by author: Lourenço Bento, 28 January 2011, Luanda; and Raul Manuel Danda, 10 February 2011, Luanda.

62. Miraldina Jamba, interview by author, 18 February 2011, Luanda.

63. Raúl Manuel Danda, interview by author, 10 February 2011, Luanda.

64. FBIS-AFR-94-010 Luanda National Radio Network, 13 January 1994. It's possible that this report drew on Danda's having recounted this story to his new colleagues at RNA.

65. Miraldina Jamba, interview by author, 18 February 2011, Luanda.

66. Miraldina Jamba, interview by author, 18 February 2011, Luanda.

67. For an account of this period, see chapter 1 in Justin Pearce, *Political Identity and Conflict in Central Angola, 1975–2002* (New York: Cambridge University Press, 2015). On UNITA members and those from the central highlands based in Luanda who fled in August 1975, see p. 37. It is likely that Dina and de Castro found themselves pushed out of Luanda as part of what Pearce calls the "territorialization" of politics during decolonization. Areas of the country and populations in those areas became associated with the politics of movements historically based there, despite individual political commitments.

68. These journalists were: Bela Malaquias, Raúl Danda, and Guida Paulo. Interviews by author: Bela Malaquias, 6 June 2008, Luanda; and Raúl Manuel Danda, 10 February 2011, Luanda.

69. Pearce, *Political Identity and Conflict*, 62.

70. Bridgland notes that the doctor he met, Adelino Manassas, was a prisoner who had been taken from Huambo in 1979. Bridgland, *Jonas Savimbi*, 322–23. See Chiwale, *Cruzei-me com a história*, 244, on teachers, religious leaders, and nurses from the missions based in the central highlands who went voluntarily.

71. Pearce, *Political Identity and Conflict*, 123–24. Pearce notes that the MPLA's political and social control in urban areas was less powerful than that of UNITA at Jamba.

MPLA-controlled cities had more diverse social organizations and were more directly heir to the complexities of colonial society.

72. Bela Malaquias, interview by author, 6 June 2008, Luanda.

73. Bela Malaquias, interview by author, 6 June 2008, Luanda.

74. See, for example, broadcasts by Voice of the Resistance of the Black Cockerel in FBIS-MEA-83-057, 23 March 1983; FBIS-MEA-83-080, 25 April 1983; FBIS-MEA-83-086, 3 May 1983; and FBIS-MEA-85-048, 12 March 1985.

75. FBIS-MEA-83-080, 25 April 1983.

76. See for example: Voice of the Resistance of the Black Cockerel broadcasts in FBIS-MEA-84-099, 21 May 1984.

77. Voice of the Resistance of the Black Cockerel, FBIS-AFR-90-026, 7 February 1990.

78. Voice of the Resistance of the Black Cockerel, FBIS-MEA-85-051, 3 April 1985.

79. Voice of the Resistance of the Black Cockerel, FBIS-MEA-85-051, 3 April 1985.

80. Voice of the Resistance of the Black Cockerel, FBIS-MEA-85-051, 3 April 1985.

81. Voice of the Resistance of the Black Cockerel, FBIS-MEA-85-051, 3 April 1985.

82. Documents attesting to this alliance emerged in the French press in 1974 and later, in 1979, in the Portuguese press in Lisbon and Luanda. William Minter, ed., *Operation Timber: Pages from the Savimbi Dossier* (Trenton, NJ: Africa World Press, 1988), 2. Minter's book brought this to the attention of the English-speaking world in the context of US funding for UNITA.

83. Voice of the Resistance of the Black Cockerel, FBIS-MEA-85-051, 3 April 1985.

84. FBIS-AFR-90-244, 19 December 1990.

85. Pearce, *Political Identity and Conflict*, chaps. 7 and 8. This was true only through the 1980s since, as Pearce notes, many Angolans had lost their confidence in this claim by the 1990s.

86. FBIS_MEA-83-086, 3 May 1983.

87. Miralinda Jamba, interview by author, 18 February 2011, Luanda.

88. Chiwale, *Cruzei-me com a história*, 247. Chiwale here recounts hearing about the 27 de Maio occurring in Luanda.

89. Miralinda Jamba, interview by author, 18 February 2011, Luanda.

90. Interviews by author: Lourenço Bento, 28 January 2011, Luanda; and Miralinda Jamba, 18 February 2011, Luanda.

91. Bela Malaquias, interview by author, 6 June 2008, Luanda.

92. Pearce *Political Identity and Conflict*. On South African support for UNITA, see Gleijeses, *Visions of Freedom*, 244–50; and William Minter, *Apartheid's Contras: An Inquiry into the Roots of War in Angola and Mozambique* (Atlantic Highlands, NJ: Zed Books, 1994), 188–91. Secret meetings between Savimbi and the South African government began in 1974, and aid was handled primarily through the military, the South African Defense Force, which had intelligence officers based at Jamba. See Gleijeses, *Visions of Freedom*, 246–47.

93. Piero Gleijeses, *Conflicting Missions: Havana, Washington, and Africa, 1959–1976* (Chapel Hill: University of North Carolina Press, 2002), 339–40.

94. Gleijeses, *Conflicting Missions*, 338–39.

95. Jardo Muekalia, *Angola—a Segunda revolução: Memórias da luta pela democracia* (Lisbon: Sextante, 2010), 54–55.

96. Gleijeses, *Conflicting Missions*, 343.

97. Chiwale, *Cruzei-me com a história*, 253–54.

98. See Pearce, *Political Identity and Conflict*, chap. 6.

99. Pearce, *Political Identity and Conflict*, 107.

100. Muekalia, *Angola*, 129.

101. Miraldina Jamba, interview by author, 18 February 2011, Luanda. Muekalia echoed this: "a base where one could think of a life very close to a normal one," in *Angola*, 129.

102. Miraldina Jamba, interview by author, 18 February 2011, Luanda; and Muekalia, *Angola*, 129–33.

103. Miraldina Jamba, interview by author, 18 February 2011, Luanda. Bridgland also visited the transport and weapons repair workshops and reports a tremendous number of vehicles and weapons captured at the battles for Mavinga in 1981. Fred Bridgland, *Jonas Savimbi*, 325.

104. Chiwale, *Cruzei-me com a história*, 242.

105. Chiwale, *Cruzei-me com a história*, 242.

106. Chiwale, *Cruzei-me com a história*, 253. Muekalia also mentions the regular army as a significant development he and colleagues noticed on their return to Jamba from Morocco in 1980. Muekalia, *Angola*, 131.

107. Bridgland, *Jonas Savimbi*, 321.

108. Bridgland, *Jonas Savimbi*, 322–23, 326–27. It's not clear whether this generator had been confiscated or came from South Africa. Chiwale notes the irony of receiving material (often military) of Eastern Bloc origin from Botha's apartheid government. Chiwale, *Cruzei-me com a história*, 250.

109. Bridgland, *Jonas Savimbi*, 323–24. Jamba's *Patriots* description of Jamba headquarters includes a telephone exchange "captured from one of the towns that had come under UNITA control." Jamba, *Patriots*, 102.

110. Bridgland, *Jonas Savimbi*, 384.

111. Radek Sikorski, "The Mystique of Savimbi," *National Review*, 18 August 1989, 36.

112. Sikorski, "Mystique of Savimbi," 36.

113. Pearce, *Political Identity and Conflict*, 111–12.

114. Tony Hodges, *Angola: Anatomy of an Oil State*, 181–82.

115. The most complete discussion is in Hodges, *Angola: Anatomy of an Oil State*, chap. 7.

116. Hodges, *Angola: Anatomy of an Oil State*, 177.

117. Hodges, *Angola: Anatomy of an Oil State*, 178–88. That said, Messiant argues that in the wake of the stillborn Lusaka Protocol, UNITA's remilitarization and diamond trading led to sanctions and official international support for the MPLA.

118. Christine Messiant, "The Mutation of Hegemonic Domination: Multiparty Politics without Democracy," in *Angola: The Weight of History*, ed. Patrick Chabal and Nuno Vidal (London: Hurst, 2007), 102–4.

119. For examples, see Voice of Resistance of the Black Cockerel, FBIS-AFR-89-154, 11 August 1989; and FBIS-AFR-89-160, 21 August 1989.

120. Elaine Windrich, *The Cold War Guerrilla: Jonas Savimbi, the U.S. Media, and the Angolan War* (New York: Greenwood Press, 1992), 117.

121. See Anita Coulson's sidebar on UNITA in "Angola: Constructing Capitalism," *Africa Report* 36, no. 6 (November 1991): 63.

122. Voice of Resistance of the Black Cockerel, FBIS-AFR-91-026, 7 February 1991.

123. Hodges, *Angola: Anatomy of an Oil State*, 13.

124. Hodges, *Angola: Anatomy of an Oil State*, 14–15.

125. For information regarding the Rapid Intervention Police, see https://www.refworld.org/docid/3df4be0110.html.

126. Pearce, *Political Identity and Conflict*, 128.

127. Bridgland, "Death in Africa," 44.

128. Amnesty International Annual Report, 1989, 33; Chiwale, *Cruzei-me com a história*, 269–71; Craig R. Whitney with Jill Jolliffe, "Ex-Allies say Angola Rebels Torture and Slay Dissenters," *New York Times*, 11 March 1989, http://www.nytimes.com/1989/03/11/world/ex-allies-say-angola-rebels-torture-and-slay-dissenters.html?pagewanted=all; Leon Dash, "Blood and Fire: Savimbi's War against His UNITA Rivals," *Washington Post*, 30 September 1990, https://www.washingtonpost.com/archive/opinions/1990/09/30/blood-and-fire-savimbis-war-against-his-unita-rivals/1e1c6ab2-e72f-45d9-aeff-2dbbd7540830/?utm_term=.0ce2e31af7e3.

129. Hodges, *Angola: Anatomy of an Oil State*, 19.

130. TPA Television Network, FBIS-AFR-92-243, 17 December 1992.

131. Amnesty International, 22 June 1992, UA 208/92—Angola: Arbitrary Detention/Fear of "Disappearance": Bela Malaquias, Germana "Tita" Malaquias, and Nelson Malaquias, index number: AFR 12/006/1992. The Index on Censorship also noted Malaquias's detention: *Index on Censorship* 21, no. 8 (1992): 34.

132. Bela Malaquias, interview by author, 6 June 2008, Luanda. Malaquias said this was a tactic Savimbi used regularly to make people "disappear quietly in the night."

133. Jardo Muekalia reflects on this incident in his memoir, *Angola*, 293–95.

134. Bela Malaquias, interview by author, 6 June 2008, Luanda; "Press Conference with Eugénio Manuvakola, Former UNITA Secretary General Regarding His Escape from UNITA," 28 August 1997, http://reliefweb.int/report/angola/press-conference-eugenio-manuvakola-former-unita-secretary-general-regarding-his; and Hodges, *Angola: Anatomy of an Oil State*, 19.

135. Ismael Mateus, "The Role of the Media during the Conflict and in the Construction of a Democracy," *Accord* 15 (2004): 63.

136. FBIS-AFR-93-066, 8 April 1993, 19: "UNITA Reproves Anstee's Behavior, 'Falsehood,'" Jamba Voz da Resistencia do Galo Negro in Portuguese, 1900 GMT, 6 April 1993. Another broadcast included in the same report announces, "UNITA says Anstee 'Morally Compromised.'"

137. FBIS-AFR-93-066, 8 April 1993, p. 19: "Observers Demand Apology for 'Insult,'" Luanda Rádio Nacional Network in Portuguese, 1900 GMT, 7 April 1993.

138. FBIS-AFR-93-068, 12 April 1993, p. 23: "UNITA Apologizes for Criticizing Anstee," London BBC World Service in English, 1515 GMT, 9 April 1993.

139. FBIS-AFR-93-068, 12 April 1993, p. 23: "UNITA Apologizes for Criticizing Anstee," London BBC World Service in English, 1515 GMT, 9 April 1993.

140. FBIS-AFR-93-068, 12 April 1993, p. 22: "UNITA to Propose End to Hostilities," Jamba Voz da Resistencia do Galo Negro in Portuguese, 0600 GMT, 10 April 1993.

141. Nick Grace C., "Investigative Report," 3.

142. FBIS-AFR-96-008, 11 January 1996, p. 35, "UNITA Warns against Attack on Jamba, Radio," Jamba Voz da Resistência do Galo Negro, 1900 GMT, 7 January 1996.

143. Notre Dame's Peace Accords Matrix charted the progress on this question and media reform under the Lusaka Protocol more generally: https://peaceaccords.nd.edu/accord/lusaka-protocol. See, too, Dawn M. Hewitt, "Peacekeeping and the Lusaka Protocol," *Melbourne Journal of Politics* 26 (1999): 87–120; and Security Council

Resolution 976 of 1995 on Angola, on the UNAVEM III UN mission radio station, http://www.un.org/Docs/scres/1995/scres95.htm, accessed 14 June 2017.

144. Protocolo de Lusaka (Governo de Angola), 49.

145. Quoted in Nick Grace C., "Investigative Report," 5.

146. Paulo de Carvalho and Reginaldo Silva, *Estado da comunicação social em Angola* (Luanda: Conselho Nacional de Comunicação Social, 2007), 14, 55.

147. Former RNA Guilherme Mogas played a critical role in this transition. Mogas, interview by author, 24 May 2011, Luanda. Nuno Vidal, "The Angolan Regime and the Move to Multiparty Politics," in Chabal and Vidal, *Angola* (London: Hurst, 2007), 148n58.

EPILOGUE: JAMMING

1. Jacques Attali, *Noise: The Political Economy of Music* (Minneapolis: University of Minnesota Press, 1985), 26. This is distinct from Larkin's use of noise as an "unintelligble signal" to colonial-era broadcasting heard but undecipherable by northern Nigerians, though offering a promise of future intelligibility. Brian Larkin, *Signal and Noise: Media, Infrastructure, and Urban Culture in Nigeria* (Durham, NC: Duke University Press, 2008), 53.

2. The MPLA's slogan in the 2012 election was "Grow more and distribute better": http://www.mpla.ao/imagem/Manifestom.pdf, accessed 16 July 2017.

3. Larkin, *Signal and Noise*, 53. Katrien Pype analyzes the constitutive role of "interference" in the "phonies" communications that connect Kinshasa to rural areas. See Pype, "On Interference and Hotspots: Ethnographic Explorations of Rural-Urban Connectivity in and Around Kinshasa's *Phonie* Cabins" (Académie Royale des Sciences d'Outre-Mer forthcoming, 2019).

4. Law 22/91, of 15 June 1991. See Paulo de Carvalho and Reginaldo Silva, *Estado da comunicação social em Angola* (Luanda: Conselho Nacional de Comunicação Social, 2007), 15.

5. Sebastião Coelho, *Angola: História e estórias da informação* (Luanda: Executive Center, 1999), 129, 137n22; and AdebayoVunge, *Dos mass mídia em Angola: Um contributo para a sua compreensão histórica* (Luanda: A. Vunge, 2006), 61.

6. Didier Péclard, "Les chemins de la 'reconversion autoritaire' en Angola," *Politique Africaine* 2, no. 110 (2008): 5–20.

7. Jorge Costa, João Teixeira Lopes, and Francisco Louçã, *Os donos angolanos de Portugal* (Lisbon: Bertrand Editora, 2014); Rafael Marques de Morais, *Diamantes de sangue: Corrupção e tortura em Angola* (Lisbon: Tinta da China, 2011); Christine Messiant, "The Mutation of Hegemonic Domination: Multiparty Politics without Democracy," in *Angola: The Weight of History*, ed. Patrick Chabal and Nuno Vidal (London: Hurst, 2007), 102–4,107–9, 113–15; Ricardo Soares de Oliveira, *Magnificent and Beggar Land: Angola since the Civil War* (London: Hurst, 2015); and Nuno Vidal, "Social Neglect and the Emergence of Angolan Civil Society," in Chabal and Vidal, *Angola*, 225.

8. Coelho, *Angola*, 130–31; Vunge, *Dos mass mídia*, 62–66.

9. Coelho, *Angola*, 131; Vunge, *Dos mass mídia*, 67–70.

10. Lúcia de Almeida, "Há cada vez mais rádios em Luanda," *Nova Gazeta*, n.d., http://novagazeta.co.ao/?p=12080, accessed 25 August 2016.

11. Almeida, "Há cada vez mais"; Milton Manaça e Ireneu Mujuco, "Radio Tocoista já emite,"*O País*, 20 July 2015, http://opais.co.ao/radio-tocoista-ja-emite/; and http://www.radioglobalfm.com/quem-somos.html.

12. Human Rights Watch, "Angola Unravels," section 8, "The Media," https://www.hrw.org/legacy/reports/1999/angola/Angl998-08.htm#P1088_190003.

13. Pacheco is the director of Associação para o Desenvolvimento Rural de Angola, (ADRA), the largest national NGO in Angola. Human Rights Watch, "Angola Unravels," section 8, "The Media," https://www.hrw.org/legacy/reports/1999/angola/Angl998-08.htm#P1088_190003.

14. Human Rights Watch, "Angola Unravels," section 8, "The Media," https://www.hrw.org/legacy/reports/1999/angola/Angl998-08.htm#P1088_190003.

15. This article points to censorship at Rádio Despertar and the pressure to broadcast UNITA's position and news: "UNITA é acusada de censurar Rádio Despertar em Angola," 15 January 2014, http://www.dw.com/pt-002/unita-%C3%A9-acusada-de-censurar-r%C3%A1dio-despertar-em-angola/a-17363626.

16. Jon Schubert, *Working the System: A Political Ethnography of the New Angola* (Ithaca, NY: Cornell University Press, 2017); and Oliveira, *Magnificent and Beggar Land*.

17. On the murder of Rádio Despertar journalist Alberto Graves Chakussanga in 2010 and the stabbing of popular radio host Manuel António "Jójó" da Silva, see Marissa J. Moorman, "Airing the Politics of Nation: Radio in Angola, Past and Present," in *Radio in Africa: Publics, Cultures, Communities*, ed. Liz Gunner, Dina Ligaga, and Dumisani Moyo (Johannesburg: Wits University Press, 2011), 238–55; on the banning of an RD journalist, see https://www.makaangola.org/2015/01/journalist-banned-from-practicing-in-cunene-province/, accessed 20 June 2017; and regarding a proposed strike by three employees in 2016, see Coque Mukuta, "Angola: Conflito no Rádio Despertar," VOA, 7 November 2016, https://www.voaportugues.com/a/angola-conflicto-na-radio-despertar/3585111.html.

18. David Lalé and Ana de Sousa, *Angola: The Birth of a Movement*, Al Jazeera, 14 October 2012, http://www.aljazeera.com/programmes/activate/2012/10/20121014131143923717.html.

19. Lalé and de Sousa, *Angola*.

20. Pedrowski Teca, "UNITA impõe censura na Rádio Despertar contra jovens revolucionários, CASA-CE e outros críticos," on Drowski blog, 24 December 2014, http://drowski3.blogspot.com/2014/12/unita-impoe-censura-na-radio-despertar.html, accessed 28 June 2017. For a more general accusation of censorship, see Human Rights Watch, "Angola Unravels," section 8, "The Media," https://www.hrw.org/legacy/reports/1999/angola/Angl998-08.htm#P1088_190003.

21. Schubert, *Working the System*.

22. John Hartley, "Radiocracy: Sound and Citizenship," *International Journal of Cultural Studies* 3, no. 2 (2000): 153–59.

23. For some examples celebrating the expanding public sphere on radio, see Tanja Bosch, "Talk Radio, Democracy and the Public Sphere: 567MW in Cape Town," in Gunner, Ligaga, and Moyo, *Radio in Africa*, 197–207; Harri Englund, *African Airwaves: Mediating Equality on the Chichewa Radio* (Bloomington: Indiana University Press, 2011); Peter Mhagama, "Radio Listening Clubs in Malawi as Alternative Public Spheres," *Radio Journal* 13, nos. 1–2 (2015): 105–20; Christopher Joseph Odhiambo, "From Diffusion to Dialogic Space: FM Radio Stations in Kenya," in Gunner, Ligaga,

and Moyo, *Radio in Africa*, 36–48. In looking at cases where there is tremendous broadcasting restriction, see Nhamo Mhiripiri, "Zimbabwe's Community Radio 'Initiatives': Promoting Alternative Media in a Restrictive Legislative Environment," *Radio Journal* 9, no. 2 (2011): 107–26; Dumisani Moyo, "Reincarnating Clandestine Radio in Post-Independent Zimbabwe," *Radio Journal* 8, no. 1 (2015): 24–26; Nicole Stremlau, Emanuele Fantini, and Iginio Gagliardone, "Patronage, Politics and Performance: Radio Call-In Programmes and the Myth of Accountability," *Third World Quarterly* 36, no. 8 (2015): 1510–26; and Wendy Willems, "Participation— In What? Radio, Convergence and the Corporate Logic of Audience Input through New Media in Zambia," *Telematics and Informatics* 30, no. 3 (2013): 223–31.

24. Wendy Willems, "Provincializing Hegemonic Histories of Media and Communications Studies: Toward a Genealogy of Epistemic Resistance in Africa," *Communication Theory* 24, no. 4 (2014): 424–25.

25. Willems, "Provincializing Hegemonic Histories," 420–22.

26. Oliveira, *Magnificent and Beggar Land*, 53–56; and Vunge, *Dos mass midia*, 55–98.

27. "Angola Unravels: The Rise and Fall of the Lusaka Peace Process," 13 September 1999, Human Rights Watch, https://www.hrw.org/legacy/reports/1999/angola/. See section 8, "The Media."

28. Péclard, "Les chemins," 12.

29. Rafael Marques de Morais, "Ecclesia: A rádio da Igreja Católica ou do governo?" on Maka Angola, 22 March 2014: https://www.makaangola.org/2014/03/ecclesia-a-radio-da-igreja-catolica-ou-do-governo/.

30. Morais, "Ecclesia: A rádio da Igreja."

31. Vunge, *Dos mass mídia*, 85–98.

32. Carvalho and Silva, *Estado da comunicação social em Angola*, 46.

33. Freedom House, Freedom of the Press Report on Angola, 2016, https://freedomhouse.org/report/freedom-press/2016/angola.

34. Vunge, *Dos mass mídia*, 97.

35. Vunge, *Dos mass mídia*, 97.

36. See this piece from Human Rights Watch: https://www.hrw.org/pt/news/2016/11/30/297073; and this one from Maka Angola https://www.makaangola.org/en/?s=Entidade%20Reguladora%20da%20Comunica%C3%A7%C3%A30%20Social%20Angolana%20ERCA.

37. Human Rights Watch, https://www.hrw.org/pt/news/2016/11/30/297073.

38. Taken from Simon Allison's 30 June 2015 article, "Reading the Revolution: The Book Club That Terrified the Angolan Regime," *Guardian*, https://www.theguardian.com/world/2015/jun/30/angola-book-club-dos-santos-arrests.

39. See Tinta da China website, http://www.tintadachina.pt/book.php?code=66685a0cec4b7dba631f30661e88a9cb.

40. For a fictitious example, see the first part of Gomes's *Tabu*, "Paraíso Perdido" (Paradise Lost). Gomes, dir., *Tabu*, DVD (Lisbon: O Som e a Fúria, 2012).

41. For examples, see the excellent short documentary by David Lalé and Ana de Sousa, *Angola: The Birth of a Movement*, Al Jazeera, 14 October 2012, http://www.aljazeera.com/programmes/activate/2012/10/20121014131143923717.html.

42. In total, seventeen people showed up. Lara Pawson, "Angola Is Stirred by the Spirit of Revolution," *Guardian*, 8 March 2011, https://www.theguardian.com/commentisfree/2011/mar/08/angola-spirit-revolution.

43. See Pawson, "Angola Is Stirred by the Spirit of Revolution," *Guardian*, 8 March 2011. But it is shortsighted to not consider the other activists and the general context that has made their emergence possible. Nuno Vidal details the constraints on as well as the growth of civil society organizations from 1991 on in Vidal, "Social Neglect."

44. From the movement's website, http://novarevolucaoangolana.yolasite.com/about-us.php, accessed 19 June 2017.

45. On Isabel dos Santos's vast wealth, see Kerry A. Dolan and Rafael Marques de Morais, "Daddy's Girl: How an African Princess Banked $3 Billion in a Country Living on $2 a Day," *Forbes*, 2 September 2013, https://www.forbes.com/sites/kerryadolan/2013/08/14/how-isabel-dos-santos-took-the-short-route-to-become-africas-richest-woman/#582292b645f5; and Forbes, https://www.forbes.com/profile/isabel-dos-santos/, accessed 29 June 2017.

46. Some in this group maintain a website on which the date of that first demonstration is seminal—7311 stands for 7 March 2011: https://centralangola7311.net/. See, too, Chloé Buire, "L'hégémonie politique à l'épreuve des musiques urbaines à Luanda, Angola," *Politique Africaine* 1, no. 141 (2016): 53–76; David Lalé and Ana de Sousa, *Angola: The Birth of a Movement*, Al Jazeera, 14 October 2012, http://www.aljazeera.com/programmes/activate/2012/10/20121014131143923717.html; and Juliana Lima, "Des 'printemps arabes' à la 'nouvelle revolucion' en Angola," *Politique Africaine* 1, no. 245 (2013): 23–36.

47. A group of Angolan artists, intellectuals, and journalists formed Liberdade Já! (Freedom Now!) and started an online campaign with YouTube videos, a Facebook presence, and letters to the press. The Facebook page is: https://www.facebook.com/Liberdade-aos-Presos-Pol%C3%ADticos-em-Angola-1606187489646481/, and their YouTube channel is https://www.youtube.com/channel/UC31vNS4gorhzW_kNBpoLqAg, accessed 29 June 2017. Amnesty International kept tabs on the situation: https://www.amnesty.org/en/countries/africa/angola/report-angola/. Front Line Defenders, a Dublin-headquartered human rights group, followed the case: https://www.frontlinedefenders.org/en/case/case-history-angola-15.

48. The interview ran on Portuguese television station SIC: https://www.youtube.com/watch?v=jQYKdv5X1zg.

49. Lima, "Des 'printemps arabes,'" 30.

50. The right to freedom of assembly, expression, and protest are guaranteed in Article 32 (1) of the Angolan constitution and by the fact of Angola's having signed on to a number of international conventions declarations, detailed here by Human Rights Watch: https://www.hrw.org/legacy/backgrounder/africa/angola/2004/4.htm, accessed 16 July 2017. Lima notes the Angolan state's nervous response in "Des 'printemps arabes,'" 24.

51. Marissa Moorman, "President dos Santos and the Ruling MPLA: Afraid of Angolans," 12 October 2015, Africasacountry.com, http://africasacountry.com/2015/10/president-dos-santos-and-the-mpla-scared-of-peace-and-quiet/; "Angola: Police Beat, Set Dogs on Peaceful Protestors," Human Rights Watch, 1 June 2017, https://www.hrw.org/news/2017/03/01/angola-police-beat-set-dogs-peaceful-protesters; "Procudora suspeita de sentir-se envergonhada nas acusações contra os revús," ClubK.net, 29 November 2015, http://club-k.net/index.php?option=com_content&view=article&id=22713:procuradora-suspeita-de-sentir-se-envergonhada-nas-acusacoes-contra-os-revus&catid=8:bastidores&Itemid=1071&lang=pt.

52. The website is Maka Angola: https://www.makaangola.org/.

53. See the Maka website: https://www.makaangola.org/about/.

54. David Lalé and Ana de Sousa, *Angola: The Birth of a Movement*, Al Jazeera, 14 October 2012, http://www.aljazeera.com/programmes/activate/2012/10/20121014131143923717.html.

55. Bertolt Brecht, "Radio as a Means of Communication: A Talk on the Function of Radio," *Screen* 20, nos. 3–4 (December 1979): 25.

56. Buire, "L'hégémonie politique"; and António Tomás, "Becoming Famous: Kuduro, Politics and the Performance of Social Visibility," *Critical Interventions* 8, no. 2 (2014): 261–75.

57. Stefanie Alisch, "Angolan *Kuduro*: *Carga*, Aesthetic Duelling, and Pleasure Politics Performed through Music and Dance" (PhD diss., Universitat Bayreuth, 2017); Buire, "L'hégémonie politique," 60; Marissa J. Moorman, "Anatomy of Kuduro: Articulating the Angolan Body Politic after the War," *African Studies Review* 57, no. 3 (2014): 21–40; Tomás, "Becoming Famous," 268.

58. Alisch, "Angolan *Kuduro*"; Buire, "L'hégémonie politique"; Moorman, "Anatomy of Kuduro"; and Tomás, "Becoming Famous."

59. See Inês Gonçalves and Kiluanje Liberdade, dirs., *Luanda: A Fábrica da Música* (Noland Filmes, 2009).

60. Buire, "L'hégémonie politique," 68. A video for the song can be found at https://www.youtube.com/watch?v=SLaj5rEzDvA.

61. On Ganga, see Shrikesh Laxmidas, "Police Kill Opposition Member, Detain 292 Protesters," Reuters, 23 November 2013, https://www.reuters.com/article/us-angola-protest-idUSBRE9AM0CZ20131123; and on the massacre at Mount Sumi, see Aslak Orre, "Covering up a Massacre in Angola?," published on the Christian Michelsen Institute website, 19 May 2015, https://www.cmi.no/news/1549-massacre.

62. See the interview on the Portuguese television station SIC: https://www.youtube.com/watch?v=jQYKdv5X1zg.

63. Lara Pawson, "Angola Is Stirred by the Spirit of Revolution," *Guardian*, 8 March 2011, cites the truculent rhetoric of MPLA leaders Bento Bento (then governor of Luanda) and Dino Matross.

64. See DW article: Pedro Borralho Ndomba, "Familiares de activistas agredidos durante marcha em Luanda," 10 August 2015, https://www.dw.com/pt-002/familiares-de-ativistas-agredidos-durante-marcha-em-luanda/a-18637857.

65. Ricardo Soares de Oliveira, "Angola's Perfect Storm: The Dos Santos Regime and the Oil Crisis," *Foreign Affairs*, 28 October 2015, https://www.foreignaffairs.com/articles/angola/2015-10-28/angolas-perfect-storm.

66. See Lima, "Des 'printemps arabes,'" 34, on the growing support of NGOs in 2011–2012 for the protesters and the boomerang effect of state repression. We can see this too in the growing number of protests organized by other groups, for example, in May and June of 2012 protests organized by the presidential guard and then by veterans of the war asking for payment of pensions in arrears. The organizers of the protest by former presidential guards Isaias Cassule and Alves Kamuliunge ended in their disappearance and death at the hands of the police.

67. At this writing, Angola's new president, João Lourenço, has been in office for less than a year. Affectionately called the implacable exonerator when he began removing dos Santos's children from positions of power in state institutions and cancelling contracts with their private companies, Angolans are now more cautious and critical of his attempts to create change. President Lourenço's prime targets were dos Santos's

children and their financial interests, which have a structuring effect on Angola's economy. After the removal of Isabel dos Santos from the directorship of Sonangol came the removal of José Filomeno "Zenu" dos Santos as director of Angola's Sovereign Wealth Fund (FSA). Less significant in terms of investments or economic effects but with profound cultural meaning, Lourenço canceled the state media's contracts with two companies owned by other dos Santos children: Welwitchia Tchize's and Coreon Dú's Semba Comunicações. Lourenço also replaced all the administrative councils of state media (television, radio, the state daily, and the newswire service).

Bibliography

Afonso, Aniceto, and Carlos de Matos Gomes, eds. *A conquista das almas: Cartazes e panlfetos da acção psicológica na guerra colonial*. Lisbon: Tinta da China, 2016.

———. *Alcora: O acordo secreto do colonialismo*. Lisbon: Objectiva, 2013.

———, eds. *Guerra colonial*. Lisbon: Editorial Notícias, 2000.

Agualusa, José Eduardo. *A teoria geral do esquecimento*. Lisbon: Dom Quixote, 2012.

———. *Estação das chuvas*. Lisbon: Dom Quixote, 2007.

Ahmed, Sara. "A Phenomenology of Whiteness." *Feminist Theory* 8, no. 2 (2007): 149–68.

Alisch, Stefanie. "Angolan *Kuduro*: *Carga*, Aesthetic Duelling, and Pleasure Politics Performed through Music and Dance." PhD diss., Universitat Bayreuth, 2017.

Allina, Eric. *Slavery by Any Other Name: African Life under Company Rule in Colonial Mozambique*. Charlottesville: University of Virginia Press, 2012.

Almeida, José Maria Pinto de. *50 anos de rádio em Angola*. Casal de Cambra: Caleidoscópio, 2016.

Alumuku, Patrick Tor. *Community Radio for Development: The World and Africa*. Nairobi: Paulines, 2006.

Amaral, Ilídio do. *Aspectos do povoamento branco de Angola*. Lisbon: Junta de Investigações do Ultramar, 1960.

———. "Contribuição para o conhecimento do fenómeno de urbanização em Angola." *Finisterra* 13, no. 25 (1978): 43–76.

———. *Luanda: Estudo de geografia urbana*. Lisbon: Memórias da Junta de Investigações do Ultramar, 1968.

Ângelo, Fernando Cavaleiro. *Os flechas: A tropa secreta da PIDE/DGS na guerra de Angola*. Alfragide: Casa das Letras, 2016.

Apostolidis, Paul. *Stations of the Cross: Adorno and Christian Right Radio*. Durham, NC: Duke University Press, 2000.

Associação Tchiweka de Documentação. *1961: Memória de um ano decisivo*. Luanda: Edições de Angola, 2015.

Attali, Jacques. *Noise: The Political Economy of Music*. Minneapolis: University of Minnesota Press, 1985.

Auerbach, Jonathan, and Russ Castronovo. "Introduction: Thirteen Propositions about Propaganda." In Auerbach and Castronovo, *Oxford Handbook of Propaganda Studies*, 1–18.

——, eds. *The Oxford Handbook of Propaganda Studies*. New York: Oxford University Press, 2013.

Baines, Gary. "The Saga of South African POWs in Angola, 1975–82." *Scientia Militaria* 40, no. 2 (2012): 102–41.

——. *South Africa's "Border War": Contested Narratives and Conflicting Memories*. London: Bloomsbury, 2014.

Baines, Gary, and Peter Vale, eds. *Beyond the Border War: New Perspectives on Southern Africa's Late Cold War Conflicts*. South Africa: UNISA Press, 2008.

Barnett, Don. "Liberation Support Movement Interview: Member of MPLA Comité Director, Daniel Chipenda, Movimento Popular de Libertação de Angola." New York: Liberation Support Movement, 1972.

——. "Liberation Support Movement Interview: Sixth Region Commander, Seta Likambuila, Movimento Popular de Libertação de Angola." New York: Liberation Support Movement, 1971.

Bastos, Mário, dir. *Independência*. Documentary. Luanda: Geração 80, 2015.

Baxter, Peter. *SAAF's Border War: The South African Air Force in Combat, 1966–1989*. Pinetown: 30 Degrees South Publishers, 2012.

Bayart, Jean-François, and Stephen Ellis. "Africa in the World: A History of Extraversion." *African Affairs* 99, no. 395 (2000): 217–67.

Beckett, Ian F. W., and John Pimlott. *Counterinsurgency: Lessons from History*. Barnsley: Pen and Sword Books, 2011.

Bender, Gerald J. *Angola under the Portuguese: The Myth and the Reality*. Berkeley: University of California Press, 1978.

Bergmeier, Horst J. P., and Rainer E. Lotz. *Hitler's Airwaves: The Inside Story of Nazi Radio Broadcasting and Propaganda Swing*. New Haven, CT: Yale University Press, 1997.

Birmingham, David. *Frontline Nationalism in Angola and Mozambique*. Trenton, NJ: Africa World Press, 1992.

——. *A Short History of Modern Angola*. London: Hurst, 2015.

——. "The Twenty-Seventh of May: An Historical Note on the Abortive 1977 'Coup' in Angola." *African Affairs* 77, no. 309 (October 1978): 554–64.

Bittencourt, Marcelo. *"Estamos juntos!": O MPLA e a luta anticolonial, 1961–1974*. 2 vols. Luanda: Kilombelombe, 2008.

Bittencourt, Marcelo, and Victor Andrade de Melo. "Esporte, economia, e política: O automobilismo em Angola (1957–1975)." *Topoi* 17, no. 32 (2016): 196–222.

Blanes, Ruy Llera. "Da confusão à ironia: Expectativas e legados da PIDE em Angola." *Análise Social* 206 (2013): 30–55.

Bloom, Peter J. "Elocution, Englishness, and Empire: Film and Radio in Late Colonial Ghana." In Bloom, Miescher, and Manuh, *Modernization as Spectacle in Africa*, 136–55.

Bloom, Peter J., Stephan F. Miescher, and Takyiwaa Manuh, eds. *Modernization as Spectacle in Africa*. Bloomington: Indiana University Press, 2014.

Bolliger, Lennart. "Apartheid's Transnational Soldiers: The Case of Black Namibian Soldiers in South Africa's Former Security Forces." *Journal of Southern African Studies* 43, no. 1 (2017): 195–214.

———. “Chiefs, Terror, and Propaganda: The Motivations of Namibian Loyalists to Fight in South African Security Force, 1975–1989.” *South African Historical Journal* 70, no. 1 (2018): 124–51.
———. “The Origins and Formation of 32 Battalion, 1975–76.” Chapter 3 of unpublished thesis, cited with author’s permission.
Bolter, Jay David, and Richard A. Grusin. *Remediation: Understanding New Media*. Cambridge, MA: MIT Press, 2000.
Bonito, Jessica Marques. “Arquitectura moderna na África Lusófona: Recepção e difusão de ideias modernas em Angola e Moçambique.” Master’s thesis, Instituto Superior Técnico, Universidade Técnica de Lisboa, 2011.
Borges, David. “A rádio de Angola estava muito mais desenvolvida do que a rádio do metropole.” In Torres, *O jornalismo Português e a guerra colonial*, 181–94.
Bosch, Tanja. “Talk Radio, Democracy and the Public Sphere: 567MW in Cape Town.” In Gunner, Ligaga, and Moyo, *Radio in Africa*, 197–207.
Botelho, Américo Cardoso. *Holocausto em Angola: Memórias de entre o cárcere e o cemitério*. Lisbon: Nova Vega, 2008.
Branch, Daniel. “Footprints in the Sand: British Colonial Counterinsurgency and the War in Iraq.” *Politics and Society* 38, no. 1 (2010): 15–34.
Branch, Daniel, and Elisabeth J. Wood. “Revisiting Counterinsurgency.” *Politics and Society* 38, no. 1 (2010): 3–14.
Brecht, Bertolt. “Radio as a Means of Communication: A Talk on the Function of Radio.” *Screen* 20, nos. 3–4 (December 1979): 24–28.
Breckenridge, Keith. “Verwoerd’s Bureau of Proof: Total Information in the Making of Apartheid.” *History Workshop Journal* 59, no. 1 (2005): 83–108.
Brennan, James R. “Communications and Media in African History.” In *The Oxford Handbook of Modern African History*, edited by John Parker and Richard Reid, 492–509. Oxford: Oxford University Press, 2013.
———. “Radio Cairo and the Decolonization of East Africa, 1953–64.” In Lee, *Making a World after Empire*, 173–95.
Bridgland, Fred. “Death in Africa.” Cold Type, 2004, coldtype.net.
———. *Jonas Savimbi: A Key to Africa*. Edinburgh: Mainstream, 1986.
Brinkman, Inge. “Refugees on Routes: Congo/Zaire and the War in Northern Angola (1961–1974).” In Heintze and von Oppen, *Angola on the Move*, 198–220.
Buire, Chloé. “L’hégémonie politique à l’épreuve des musiques urbaines à Luanda, Angola.” *Politique Africaine* 1, no. 141 (2016): 53–76.
Burchett, Wilfred G., and Derek Roebuck. *The Whores of War: Mercenaries Today*. London: Penguin, 1977.
Cann, John P. *Counterinsurgency in Africa: The Portuguese Way of War, 1961–1974*. Pinetown: Helion, 2013.
———. *The Flechas: Insurgent Hunting in Eastern Angola, 1965–1974*. Pinetown: Helion, 2013.
Cardoso, António. *Baixa e musseques*. Havana: Ediciones Cubanas for the União de Escritores Angolanos, 1985.
Cardoso, Dulce Maria. *O retorno*. Lisbon: Tinta da China, 2012.
Carvalho, Paulo de. *Audiência de media em Luanda*. Luanda: Coleção Ensaio, 2002.
Carvalho, Paulo de, and Reginaldo Silva. *Estado da comunicação social em Angola*. Luanda: Conselho Nacional de Comunicação Social, 2007.

Castelo, Cláudia. *"O modo português de estar no mundo": O luso-tropicalismo e a ideológica colonial portuguesa (1933–1961)*. Porto: Edições Afrontamento, 1998.

———. *Passagens para África: O povoamento de Angola e Moçambique com naturais da metrópole (1920–1974)*. Porto: Edições Afrontamento, 2007.

Cervelló, Joseph Sanchez. "Caso Angola." In Afonso and Gomes, *Guerra colonial*.

Chabal, Patrick, and Nuno Vidal, eds. *Angola: The Weight of History*. London: Hurst, 2007.

Chikowero, Mhoze. "Is Propaganda Modernity? Press and Radio for 'Africans' in Zambia, Zimbabwe, and Malawi during World War II and Its Aftermath." In Bloom, Miescher, and Manuh, *Modernization as Spectacle in Africa*, 112–35.

Chiwale, Samuel. *Cruzei-me com a história*. Lisbon: Sextante Editora, 2008.

Clarence-Smith, Gervase. *The Third Portuguese Empire, 1825–1975: A Study in Economic Imperialism*. Manchester: Manchester University Press, 1985.

Cleveland, Todd. "'We Still Want the Truth': The ANC's Angolan Detention Camps and Post-Apartheid Memory." *Comparative Studies of South Asia, Africa and the Middle East* 25, no. 1 (2005): 63–78.

Coelho, Sebastião. *Angola: História e estórias da informação*. Luanda: Executive Center, 1999.

Collier, Delinda. "A 'New Man' for Africa? Some Particularities of the Marxist *Homem Novo* within Angolan Cultural Policy." In *De-Centering Cold War History: Local and Global Change*, edited by Jadwiga E. Pieper Mooney and Fabio Lanza, 187–206. New York: Routledge, 2013.

———. *Repainting the Walls of Lunda: Information Colonialism and Angolan Art*. Minneapolis: University of Minnesota Press, 2016.

Comaroff, Jean, and John L. Comaroff. *Of Revelation and Revolution*. Vol. 1, *Christianity, Colonialism, and Consciousness in South Africa*. Chicago: University of Chicago Press, 1991.

———. *Of Revelation and Revolution*. Vol. 2, *The Dialectics of Modernity on a South African Frontier*. Chicago: University of Chicago Press, 1997.

Conchiglia, Augusta. *Guerra di popolo in Angola*. Rome: Lerici, 1969.

Connelly, Matthew. *A Diplomatic Revolution: Algeria's Fight for Independence and the Origins of the Post–Cold War Era*. Oxford: Oxford University Press, 2003.

———. "Rethinking the Cold War and Decolonization: The Grand Strategy of the Algerian War for Independence." *International Journal of Middle East Studies* 33, no. 2 (2001): 221–45.

Cooper, Frederick. *Decolonization and African Society: The Labor Question in French and British Africa*. Cambridge: Cambridge University Press, 1996.

Cooper, Frederick, and Ann Laura Stoler, eds. *Tensions of Empire: Colonial Cultures in a Bourgeois World*. Berkeley: University of California Press, 1997.

Coplan, David B. "South African Radio in a Saucepan." In Gunner, Ligaga, and Moyo, *Radio in Africa*, 134–48.

Costa, Jorge, João Teixeira Lopes, and Francisco Louçã. *Os donos angolanos de Portugal*. Lisbon: Bertrand Editora, 2014.

Coulson, Anita. "Angola: Constructing Capitalism." *Africa Report* 36, no. 6 (November 1991): 61–65.

Craig, Douglas B. "Radio, Modern Communication Media and the Technical Sublime." *Radio Journal* 6, nos. 2–3 (2008): 129–43.

Cruz, Gabriela. "The Suspended Voice of Amália Rodrigues." In *Music in Print and Beyond: Hildegard von Bingen to the Beatles*, edited by Craig A. Monson and Roberta Montemorra Marvin, 180–99. Rochester, NY: University of Rochester Press, 2013.

Cull, Nicholas J. "Roof for a House Divided: How U.S. Propaganda Evolved into Public Diplomacy." In Auerbach and Castronovo, *Oxford Handbook of Propaganda Studies*, 131–46.

Cunha, Anabela. "'Processo dos 50': Memórias de luta clandestina pela independência de Angola." *Revista Angolana de Sociologia* 8 (2011): 87–96.

Darling, Juanita. "Radio and Revolution in El Salvador: Building a Community of Listeners in the Midst of Civil War, 1981–1992." *American Journalism* 24, no. 4 (2007): 67–93.

Dáskalos, Sócrates. *Um testemunho para a história de Angola: Do Huambo ao Huambo*. Lisbon: Vega, 2000.

Davidson, Basil. *In the Eye of the Storm: Angola's People*. New York: Doubleday, 1972.

Davis, Stephen R. 'The African National Congress, Its Radio, Its Allies and Exile." *Journal of Southern African Studies* 35, no. 2 (2009): 349–73.

Day, Christopher R., and William S. Reno. "In Harm's Way: African Counter-Insurgency and Patronage Politics." *Civil Wars* 16, no. 2 (2014): 105–26.

De Grassi, Aharon. "Provisional Reconstructions: Geo-Histories of Infrastructure and Agrarian Configuration in Malanje, Angola." PhD diss., University of California, Berkeley, 2015.

———. "Rethinking the 1961 Baixa de Kassanje Revolt: Towards a Relational Geo-History of Angola." *Mulemba* 5, no. 10 (2015): 53–103.

De Kok, Ingrid. "The Sound Engineer." In *Seasonal Fires: New and Selected Poems*. Houghton: Umuzi, 2006.

Douglas, Susan J. *Listening In: Radio and the American Imagination*. Minneapolis: University of Minnesota Press, 1999.

Edwards, Paul N., and Gabrielle Hecht. "History and the Technopolitics of Identity: The Case of Apartheid South Africa." *Journal of Southern African Studies* 36, no. 3 (September 2010): 619–39.

Elkins, Caroline, and Susan Pedersen. "Introduction: Settler Colonialism; a Concept and Its Uses." In Elkins and Pedersen, *Settler Colonialism in the Twentieth Century*, 1–20.

———, eds. *Settler Colonialism in the Twentieth Century: Projects, Practices, Legacies*. New York: Routledge, 2005.

Ellis, Stephen. "The ANC in Exile." *African Affairs* 90, no. 360 (1991): 439–47.

———"Tuning in to Pavement Radio." *African Affairs* 88, no. 352 (1989): 321–30.

Englund, Harri. *Human Rights and African Airwaves: Mediating Equality on the Chichewa Radio*. Bloomington: Indiana University Press, 2011.

Esteves, Emmanuel. "As vias de comunicação e meios de transporte como factores de globalizacação, de estabilidade política e de transformação económica e social: Caso do Caminho-de-Ferro de Bengela (Benguela) (1889–1950)." In Heintze and von Oppen, *Angola on the Move*, 99–116.

Fanon, Frantz. A *Dying Colonialism*. Translated by Haakon Chevalier. New York: Grove Press, 1965.

Fardon, Richard, and Graham Furniss, eds. *African Broadcast Cultures: Radio in Transition*. London: James Currey, 2000.

Fernandes, Jason Keith. "A Goan Waltz around Post-Colonial Dogmas." Unpublished paper from graduate workshop "Europe and the World: Between Colonialism and Globalization," Villa Vigoni, 18–22 June 2018.
Ferreira, Manuel Ennes. *A indústria em tempo de guerra (Angola, 1975–91)*. Lisbon: Edições Cosmos, 1999.
———. "Nacionalização e confisco do capital português na indústria transformadora de Angola (1975–1990)." *Análise Social* 37, no. 162 (2002): 47–90.
Filipe, Eléusio dos Prazeres Viegas. "Voice of FRELIMO Insurrection against the Voice of Mozambique: Waging Propaganda War against Portuguese Colonialism and Building a National Consciousness, 1960s–1975." Paper presented at the Wits Workshop on Liberation Radios in Southern Africa, University of the Witswatersrand, February 2017.
Fisher, Daniel. "Radio." In Novak and Sakakeeny, *Keywords in Sound*, 151–64.
Fiúza, Filipa, and Ana Vaz Milheiro. "The Prenda District in Luanda: Building on Top of the Colonial City." In *Urban Planning in Lusophone African Countries*, edited by Carlos Nunes Silva, 93–100. Surrey: Ashgate, 2015.
Flores, Paulo. *Bolo de aniversário*. CD. Lisbon: Frequento Aplauso, 2016.
Fonseca, Ana Sofia. *Angola, terra prometida: A vida que os portugueses deixaram*. Lisbon: Esfera dos Livros, 2009.
Francisco, Miguel "Michel." *Nuvem negra: O drama do 27 de Maio de 1977*. Lisbon: Clássica Editora, 2007.
Freudenthal, Aida. "A Baixa de Cassanje: Algodão e revolta." *Revista Internacional de Estudos Africanos* 18–22 (1995–1999): 245–83.
———. "Baixa de Kassanje, 1961." In Associação Tchiweka de Documentação, *1961: Memória de um ano decisivo*, 13–16.
Gamboa, Zézé, dir. *O grande kilapy*. DVD. Lisbon: David and Golias, 2012.
Garcia, Rita. *Luanda como ela era, 1960–1975: Histórias e memórias de uma cidade inesquecível. Alfragide*: Oficina do Livro, 2016.
Gastrow, Claudia. "Negotiated Settlements: Housing and the Aesthetics of Citizenship in Luanda, Angola." PhD diss., University of Chicago, 2014.
Geldenhuys, Jannie. *A General's Story: From an Era of War and Peace*. Johannesburg: Jonathan Ball, 1995.
George, Edward. *The Cuban Intervention in Angola, 1965–1991: From Che Guevara to Cuito Cuanavale*. New York: Frank Cass, 2005.
Glaser, Clive. "The Making of a Portuguese Community in South Africa, (1900–1994)." In Morier-Genoud and Cahen, *Imperial Migrations*, 213–38.
Gleijeses, Piero. *Conflicting Missions: Havana, Washington, and Africa, 1959–1976*. Chapel Hill: University of North Carolina Press, 2002.
———. *Visions of Freedom: Havana, Washington, Pretoria, and the Struggle for Southern Africa, 1976–1991*. Chapel Hill: University of North Carolina Press, 2013.
Gomes, Carlos de Matos. "A africanização na guerra colonial e as suas sequelas tropas locais: Os vilões nos ventos da história." In Meneses and Martins, *As guerras da libertação e os sonhos coloniais*, 123–41.
Gomes, Miguel, dir. *Tabu*. DVD. Lisbon: O Som e a Fúria, 2012.
Gonçalves, Inês, and Kiluanje Liberdade, dirs. *Luanda: A Fábrica da Música*. Noland Films, 2009.
Gondola, Ch. Didier. *The History of the Congo*. Westport, CT: Greenwood Press, 2002.

Goscha, Christopher. "Wiring Decolonization: Turning Technology against the Colonizer during the Indochina War, 1945–1954." *Comparative Studies in Society and History* 54, no. 4 (2012): 798–831.

Gray, Lila Ellen. *Fado Resounding: Affective Politics and Urban Life*. Durham, NC: Duke University Press, 2013.

Gunner, Liz. "Wrestling with the Present, Beckoning to the Past: Contemporary Zulu Radio Drama." *Journal of Southern African Studies* 26, no. 2 (2000): 223–37.

Gunner, Liz, Dina Ligaga, and Dumisani Moyo, eds. *Radio in Africa: Publics, Cultures, Communities*. Johannesburg: Wits University Press, 2011.

Gupta, Pamila. "Decolonization and (Dis)Possession in Lusophone Africa." In *Mobility Makes States: Migration and Power in Africa*, edited by Darshan Vigneswaran and Joel Quirk, 169–93. Philadelphia: University of Pennsylvania Press, 2015.

——. "'Going for a Sunday Drive': Angolan Decolonization, Learning Whiteness and the Portuguese Diaspora of South Africa." In *Narrating the Portuguese Diaspora: Piecing Things Together*, edited by Francisco Cota Fagundes, Irene Maria F. Blayer, Teresa F. A. Alves, and Teresa Cid, 135–52. New York: Peter Lang, 2011.

Hartley, John. "Radiocracy: Sound and Citizenship." *International Journal of Cultural Studies* 3, no. 2 (2000): 153–59.

Hecht, Gabrielle. *Being Nuclear: Africans and the Global Uranium Trade*. Cambridge, MA: MIT Press, 2012.

——, ed. *Entangled Geographies: Empire and Technopolitics in the Cold War*. Cambridge, MA: MIT Press, 2011.

Heidegger, Martin. *The Question Concerning Technology and Other Essays*. Translated by William Lovitt. New York: Garland, 1977.

Heintze, Beatrix, and Achim von Oppen, eds. *Angola on the Move: Transport Routes, Communications and History/Angola em Movimento: Vias de Transporte, Comunicação e História*. Frankfurt am Main: Lembeck, 2008.

Heinze, Robert. "'It Recharged Our Batteries': Writing the History of the Voice of Namibia." *Journal of Namibian Studies* 15 (2014): 25–62.

Helmreich, Stefan. "Transduction." In Novak and Sakakeeny, *Keywords in Sound*, 222–31.

Herbst, Jeffrey. *States and Power in Africa: Comparative Lessons in Authority and Control*. Princeton, NJ: Princeton University Press, 2014.

Herf, Jeffrey. "Narrative and Mendacity: Anti-Semitic Propaganda in Nazi Germany." In Auerbach and Castronovo, *Oxford Handbook of Propaganda Studies*, 91–108.

Hewitt, Dawn M. "Peacekeeping and the Lusaka Protocol." *Melbourne Journal of Politics* 26 (1999): 87–120.

Heywood, Linda M., and John K. Thornton. *Central Africans, Atlantic Creoles, and the Foundation of the Americas, 1585–1660*. New York: Cambridge University Press, 2007.

Hilmes, Michele. *Radio Voices: American Broadcasting, 1922–1952*. Minneapolis: University of Minnesota Press, 1997.

Hodges, Tony. *Angola: Anatomy of an Oil State*. Bloomington: Indiana University Press, 2001. First published as *Angola: From Afro-Stalinism to Petro-Diamond Capitalism* (Bloomington: Indiana University Press, 2001).

Hunt, Nancy Rose. "An Acoustic Register: Rape and Repetition in Congo." In *Imperial Debris: On Ruins and Ruination*, edited by Ann Laura Stoler, 39–66. Durham, NC: Duke University Press, 2013.

——. *A Nervous State: Violence, Remedies, and Reverie in Colonial Congo*. Durham, NC: Duke University Press, 2016.
Isaacman, Allen F. *Cotton Is the Mother of Poverty: Peasants, Work, and Rural Struggle in Colonial Mozambique, 1938–1961*. Portsmouth, NH: Heinemann, 1996.
Isaacman, Allen F., and Barbara S. Isaacman. *Dams, Displacement, and the Delusion of Development: Cahora Bassa and Its Legacies in Mozambique, 1965–2007*. Athens: Ohio University Press, 2013.
——. *Mozambique: From Colonialism to Revolution, 1900–1982*. Boulder, CO: Westview Press, 1993.
Jamba, Sousa. *Patriots*. London: Viking, 1990.
James, W. Martin, III. *A Political History of the Civil War in Angola, 1974–1990*. New Brunswick, NJ: Transaction Publishers, 2011.
Jerónimo, Miguel Bandeira, and António Costa Pinto. "A Modernizing Empire? Politics, Culture, and Economy in Portuguese Late Colonialism." In *Ends of European Colonial Empires*, 51–80.
——, eds. *The Ends of European Colonial Empires: Cases and Comparisons*. New York: Palgrave Macmillan, 2015.
Júnior, Miguel. *A mão Sul-Africana: O envolvimento das forças de defesa da África do Sul no sudeste de Angola (1966–1974)*. Caiscais: Tribuna de História, 2014.
Kabir, Ananya Jahanara. "The Dancing Couple in Black Atlantic Space." In *Diasporic Women's Writing of the Black Atlantic: (En)Gendering Literature and Performance*, edited by Emilia María Durán-Almarza and Esther Álvarez López, 133–50. New York: Routledge, 2013.
——. "Oceans, Cities, Islands: Sites and Routes of Afro-Diasporic Rhythm Cultures." *Atlantic Studies* 11, no. 1 (2014): 106–24.
Kennedy, Dane K. *Islands of White: Settler Society and Culture in Kenya and Southern Rhodesia, 1890–1939*. Durham, NC: Duke University Press, 1987.
Kennes, Erik, and Miles Larmer. *The Katangese Gendarmes and War in Central Africa: Fighting Their Way Home*. Bloomington: Indiana University Press, 2016.
Kushner, James M. "African Liberation Broadcasting." *Journal of Broadcasting* 18, no. 3 (1974): 299–310.
Labuschagne, Bernice. "South Africa's Intervention in Angola: Before Cuito Cuanavale and Thereafter." Master's thesis, Stellenbosch University, 2009.
Landau, Paul. "Empires of the Visual: Photography and Colonial Administration in Africa." In *Images and Empires: Visuality in Colonial and Postcolonial Africa*, edited by Paul S. Landau and Deborah D. Kaspin, 141–71. Berkeley: University of California Press, 2002.
——. *The Realm of the Word: Language, Gender and Christianity in a Southern African Kingdom*. Portsmouth, NH: Heinemann, 1995.
Larkin, Brian. *Signal and Noise: Media, Infrastructure, and Urban Culture in Nigeria*. Durham, NC: Duke University Press, 2008.
Lee, Christopher J. "Decolonization of a Special Type: Rethinking Cold War History in Southern Africa." *Kronos* 37, no. 1 (2011): 6–11.
——, ed. *Making a World after Empire: The Bandung Moment and Its Political Afterlives*. Athens: Ohio University Press, 2010.
Lekgoathi, Sekibakiba Peter. "The African National Congress's Radio Freedom and Its Audiences in Apartheid South Africa, 1963–1991." *Journal of African Media Studies* 2, no. 2 (2010): 139–53.

———. "Bantustan Identity, Censorship and Subversion on Northern Sotho Radio under Apartheid, 1960s–80s." In Gunner, Ligaga, and Moyo, *Radio in Africa*, 117–33.

Lima, Juliana. "Des 'printemps arabes' à la 'nouvelle revolucion' en Angola." *Politique Africaine* 1, no. 245 (2013): 23–36.

Lopes, Júlio Mendes. "A evolução da radiodifusão em Angola e a problemática da difusão da história da Àfrica." PhD diss., Universidade de Agostinho Neto, Instituto Superior de Ciências de Educação, Luanda, 1997.

Loviglio, Jason. *Radio's Intimate Public: Network Broadcasting and Mass-Mediated Democracy*. Minneapolis: University of Minnesota Press, 2005.

Lubkemann, Stephen C. "Unsettling the Metropole: Race and Settler Reincorporation in Postcolonial Portugal." In Elkins and Pedersen, *Settler Colonialism*, 257–70.

MacQueen, Norrie. *The Decolonization of Portuguese Africa: Metropolitan Revolution and the Dissolution of Empire*. New York: Longman, 1997.

Magalhães, Ana. *Moderno tropical: Arquitectura em Angola e Moçambique 1948–1975*. Lisbon: Tinta da China, 2009.

Maier, Karl. *Angola: Promises and Lies*. Rivonia: William Waterman, 1996.

Malaquias, Assis. *Rebels and Robbers: Violence in Post-colonial Angola*. Uppsala: Nordic Africa Institute, 2007.

Manyozo, Linje. "Mobilizing Rural and Community Radio in Africa." *Ecquid Novi* 30, no. 1 (2009): 1–23.

Marcum, John A. *The Anatomy of an Explosion (1950–1962)*. Vol. 1 of *The Angolan Revolution*. Cambridge, MA: MIT Press, 1969.

Marx, Leo. *The Machine in the Garden: Technology and the Pastoral Ideal in America*. 35th anniversary ed. Oxford: Oxford University Press, 2000.

Mateus, Dalila Cabrita. *A PIDE/DGS na guerra colonial, 1961–1974*. Lisbon: Terramar, 2004.

Mateus, Dalila Cabrita, and Álvaro Mateus. *Angola '61 guerra colonial: Causas e consequências*. Lisbon: Almedina, 2011.

———. *Purga em Angola: Nito Alves, Sita Valles, Zé Van Dunem, o 27 de Maio de 1977*. Lisbon: Edições ASA, 2007.

Mateus, Ismael. "The Role of the Media during the Conflict and in the Construction of a Democracy." *Accord* 15 (2004): 62–65.

Mavhunga, Clapperton. "A Plundering Tiger with Its Deadly Cubs? The USSR and China as Weapons in the Engineering of a 'Zimbabwean Nation,' 1945–2009." In Hecht, *Entangled Geographies*, 231–66.

———. "Usage Engineers as a Force behind the Movement of Technology, Part 1: Guns in Pre-Colonial Africa." Paper presented at Department of Science and Technology Studies Seminar Series, Cornell University, September 2012. Cited with author's permission.

McCaskie, Tom. "Unspeakable Words, Unmasterable Feelings: Calamity and the Making of History in Asante." *Journal of African History* 59, no. 1 (2018): 3–20.

Medeiros, Carlos Alberto. *A colonização das terras altas da Huíla (Angola)*. Lisbon: Centro de Estudos Geográficos, 1976.

Megwa, Eronini R. "Bridging the Digital Divide: Community Radio's Potential for Extending Information and Communication Technology Benefits to Poor Rural Communities in South Africa." *Howard Journal of Communications* 18, no. 4 (2007): 335–52.

Meichenbaum, Donald. "Helping the Helpers." In *Trauma and Post-Traumatic Stress Disorder*, edited by Michael J. Scott and Stephen Palmer, 117–21. London: Sage, 2000.
Meneses, Maria Paula, and Catarina Gomes. "Regressos? Os *retornados* na (des) colonização portuguesa." In Meneses and Martins, *As guerras de libertação e os sonhos coloniais*, 59–108.
Meneses, Maria Paula, and Bruno Sena Martins, eds. *As guerras de libertação e os sonhos coloniais: Alianças secretas, mapas imaginados*. Lisbon: Almedina, 2013.
Meneses, Maria Paula, Celso Braga Rosa, and Bruno Sena Martins. "Estilhaços do *Exercício Alcora*: O epílogo dos sonhos coloniais." In Meneses and Martins, *As guerras de libertação e os sonhos coloniais*, 171–79.
Messiant, Christine. *L'Angola colonial, histoire et société: Les prémisses du mouvement nationaliste*. Switzerland: P. Schlettwein, 2006.
———. "The Mutation of Hegemonic Domination: Multiparty Politics without Democracy." In Chabal and Vidal, *Angola*, 93–123.
Mhagama, Peter. "Radio Listening Clubs in Malawi as Alternative Public Spheres." *Radio Journal* 13, nos. 1–2 (2015): 105–20.
Mhiripiri, Nhamo. "Zimbabwe's Community Radio 'Initiatives': Promoting Alternative Media in a Restrictive Legislative Environment." *Radio Journal* 9, no. 2 (2011): 107–26.
Mhlambi, Thokozani N. "Early Radio Broadcasting in South Africa: Culture, Modernity and Technology." PhD diss., University of Cape Town, South African College of Music, 2015.
Milhazes, José. *Angola: O Princípio do fim da União Soviética*. Lisbon: Vega, 2009.
Mills, Mara. "Deafness." In Novak and Sakakeeny, *Keywords in Sound Studies*, 45–54.
Minter, William. *Apartheid's Contras: An Inquiry into the Roots of War in Angola and Mozambique*. Atlantic Highlands, NJ: Zed Books, 1994.
———, ed. *Operation Timber: Pages from the Savimbi Dossier*. Trenton, NJ: Africa World Press, 1988.
Monteiro, Diamantino Pereira. "A rádio de Angola era um prolongamento da rádio portuguesa." In Torres, *O jornalismo português e a guerra colonial*, 173–80.
———. *Rádio em Angola: Como eu a vivi*. Coimbra: Mar da Palavra, 2018.
Monteiro, Ramiro Ladeiro. "Subsídios para a história recente das informações em Portugal." In Moreira, *Informações e segurança*, 459–70.
Moorman, Marissa J. "Airing the Politics of Nation: Radio in Angola, Past and Present." In Gunner, Ligaga, and Moyo, *Radio in Africa*, 238–55.
———. "Anatomy of Kuduro: Articulating the Angolan Body Politic after the War." *African Studies Review* 57, no. 3 (2014): 21–40.
———. *Intonations: A Social History of Music and Nation in Luanda, Angola, from 1945 to Recent Times*. Athens: Ohio University Press, 2008.
———. "Putting on a *Pano* and Dancing Like Our Grandparents: Nation and Dress in Late Colonial Luanda." In *Fashioning Africa: Power and the Politics of Dress*, edited by Jean Allman, 84–103. Bloomington: Indiana University Press, 2004.
———. "Radio Remediated: Sissako's *Life on Earth* and Sembène's *Moolaadé*." *Cinema Journal* 57, no. 1 (2017): 94–116.
Morais, Rafael Marques de. *Diamantes de sangue: Corrupção e tortura em Angola*. Lisbon: Tinta da China, 2011.
Moreira, Adriano, ed. *Informações e segurança: Estudos em honra de General Pedro Cardoso*. Lisbon: Prefácio, 2004.

Morier-Genoud, Eric, and Michel Cahen, eds. *Imperial Migrations: Colonial Communities and Diaspora in the Portuguese World*. New York: Palgrave Macmillan, 2012.
Movimento Popular para a Libertação de Angola–Partido do Trabalho (MPLA-PT). "Teses e resoluções, 1° Congresso." Luanda: Imprensa Nacional, 1978.
Mowitt, John. *Radio: Essays in Bad Reception*. Berkeley: University of California Press, 2011.
Moyo, Dumisani. "Reincarnating Clandestine Radio in Post-Independent Zimbabwe." *Radio Journal* 8, no. 1 (2015): 23–36.
Moyo, Dumisani, and Cris Chinaka. "Persuasion, Propaganda and Mass Mobilisation through Underground Radio in the Zimbabwe War of Liberation: A Comparative Study of Radio Voice of the People and Voice of the Revolution." Paper circulated at the Wits Worskhop on Liberation Radios in Southern Africa, University of the Witwatersrand, February 2017.
MPLA-PT. *See* Movimento Popular para a Libertação de Angola–Partido do Trabalho.
Muekalia, Jardo. *Angola—a Segunda revolução: Memórias da luta pela democracia*. Lisbon: Sextante, 2010.
Murnau, F. W., dir. *Tabu*. Paramount, 1931.
Myers, Mary. "Community Radio and Development: Issues and Examples from Francophone West Africa." In Fardon and Furniss, *African Broadcast Cultures*, 90–101.
Mytton, Graham. "From Saucepan to Dish: Radio and TV in Africa." In Fardon and Furniss, *African Broadcast Cultures*, 21–41.
———. *Mass Communication in Africa*. London: Edward Arnold, 1983.
Neto, Maria da Conceição. "In Town and Out of Town: A Social History of Huambo (Angola), 1902–1961." PhD thesis, SOAS, University of London, 2012.
———. "Nas malhas da rede: Aspectos do impacto económico e social do transporte rodoviário na região do Huambo c. 1920–c. 1960." In Heintze and von Oppen, *Angola on the Move*, 117–29.
Novak, David, and Matt Sakakeeny, eds. *Keywords in Sound*. Durham, NC: Duke University Press, 2015.
Nzongola-Ntalaja, Georges. *The Congo from Leopold to Kabila: A People's History*. New York: Zed Books 2002.
Odhiambo, Christopher Joseph. "From Diffusion to Dialogic Space: FM Radio Stations in Kenya." In Gunner, Ligaga, and Moyo, *Radio in Africa*, 36–48.
Oliveira, Pedro Aires. "Live and Let Live: Britain and Portugal's Imperial Endgame (1945–75)." *Portuguese Studies* 29, no. 2 (2013): 186–208.
Oliveira, Ricardo Soares de. "Business Success, Angola-Style: Postcolonial Politics and the Rise and Rise of Sonangol." *Journal of Modern African Studies* 45, no. 4 (2007): 595–619.
———. *Magnificent and Beggar Land: Angola since the Civil War*. London: Hurst, 2015.
Olorunnisola, Anthony A. "Community Radio: Participatory Communication in Postapartheid South Africa." *Journal of Radio Studies* 9, no. 1 (2002): 126–45.
Ondjaki. *Good Morning Comrades: A Novel*. Translated by Stephen Henighan. Emeryville: Biblioasis, 2001. Originally published as *Bom dia camaradas: Romance* (Luanda: Chá de Caxinde, 2000).
———. *Os transparentes*. Lisbon: Caminho, 2012.
Onslow, Sue, ed. *Cold War in Southern Africa: White Power, Black Liberation*. New York: Routledge, 2009.

———. "The Cold War in Southern Africa: White Power, Black Nationalism and External Intervention." In Onslow, *Cold War in Southern Africa*, 9–34.
Opoku-Mensah, Aida. "The Future of Community Radio in Africa." In Fardon and Furniss, *African Broadcast Cultures*, 165–73.
Paredes, Margarida Isabel Botelho Falcão. "Mulheres na luta armada em Angola: Memória, cultura e emancipação." PhD diss., ISCTE–Instituto Universitário de Lisboa, 2014.
Pawson, Lara. *In the Name of the People: Angola's Forgotten Massacre*. New York: Tauris, 2014.
Pearce, Justin. *Political Identity and Conflict in Central Angola, 1975–2002*. New York: Cambridge University Press, 2015.
Péclard, Didier. "Les chemins de la 'reconversion autoritaire' en Angola." *Politique Africaine* 2, no. 110 (2008): 5–20.
———. *Les incertitudes de la nation en Angola: Aux racines sociales de l'Unita*. Paris: Karthala, 2015.
Pélissier, René. *La colonie du Minotaure: Nationalismes et revoltes en Angola, 1926–1961*. Orgeval: Pélissier, 1978.
Penvenne, Jeanne Marie. *African Workers and Colonial Racism: Mozambican Strategies and Struggles in Lourenço Marques, 1877–1962*. Portsmouth, NH: Heinemann, 1994.
———. "Settling against the Tide: The Layered Contradictions of Twentieth-Century Portuguese Settlement in Mozambique." In Elkins and Pedersen, *Settler Colonialism in the Twentieth Century*, 79–94.
Pepetela. *A Geração da Utopia*. Lisbon: Dom Quixote, 1997.
Peters, Christabelle. *Cuban Identity and the Angolan Experience*. New York: Palgrave Macmillan, 2012.
Peters, John Durham. "Helmholtz, Edison, and Sound History." In *Memory Bytes: History, Technology, and Digital Culture*, edited by Lauren Rabinovitz and Abraham Geil, 177–98. Durham, NC: Duke University Press, 2004.
———. *Speaking into the Air: A History of the Idea of Communication*. Chicago: University of Chicago Press, 2000.
Peterson, Derek R. *Creative Writing: Translation, Bookkeeping, and the Work of Imagination in Colonial Kenya*. Portsmouth, NH: Heinemann, 2004.
———. *Ethnic Patriotism and the East African Revival: A History of Dissent, c. 1935–1972*. Cambridge: Cambridge University Press, 2014.
Pimenta, Fernando Tavares. "A ditadura e a emergência do nacionalismo entre os brancos de Angola." In *Angola, os brancos e a indepêndencia*, 137–216.
———. *Angola, os brancos e a independência*. Lisbon: Edições Afrontamento, 2008.
———. *Angola no percurso de um nacionalista: Conversas com Adolfo Maria*. Lisbon: Edições Afrontamento, 2006.
———. "Ideologia nacional dos brancos angolanos." In *Identidades, memórias e histórias, em terras africanas*, edited by Selma Pantoja, 169–91. Brasilia: LGE, 2006.
———. "O estado novo português e a reforma do estado colonial em Angola: O comportamento político das elites brancas (1961–1962)." *História* 33, no. 2 (2014): 250–72.
Pimentel, Irene Flunser. *A história da PIDE*. Lisbon: Círculo de Leitores, 2011.
Pinto, José Filipe. "A censura em Angola durante a guerra colonial." In Torres, *O jornalismo Português e a guerra colonial*, 121–27.
Pinto, Renato Marques. "Os militares e as informações (em memória do General Pedro Cardoso)." In Moreira, *Informações e segurança*, 471–89.

Pinto, Tatiana Pereira Leite. "Etnicidade, racismo e luta em Angola: As questões étnicas e raciais na luta de libertação e no governo de Agostinho Neto." BA diss., Universidade Federal Fluminense, 2008.
Pitcher, M. Anne. *Transforming Mozambique: The Politics of Privatization, 1975–2000*. Cambridge: Cambridge University Press, 2002.
Pombo [Harry Villegas]. *Cuba and Angola: The War for Freedom*. New York: Pathfinder, 2017.
Posel, Deborah. "The Case for a Welfare State: Poverty and the Politics of the Urban African Family in the 1930s and 1940s." In *South Africa's 1940s: Worlds of Possibilities*, edited by Saul Dubow and Alan Jeeves, 64–86. Cape Town: Double Storey, 2005.
Prashad, Vijay. *The Darker Nations: A People's History of the Third World*. New York: New Press, 2008.
Pratt, Frederick. "'Ghana Muntie!': Broadcasting, Nation-Building, and Social Difference in the Gold Coast and Ghana, 1935–1985." PhD diss., Indiana University, 2013.
Prendergast, Maria Teresa, and Thomas A. Prendergast. "The Invention of Propaganda: A Critical Commentary on and Translation of *Inscrutabili Divinae Providentiae Arcano*." In Auerbach and Castronovo, *Oxford Handbook of Propaganda Studies*, 19–27.
Pype, Katrina. "On Interference and Hotspots: Ethnographic Explorations of Rural-Urban Connectivity in and Around Kinshasa's Phonie Cabins." Académie Royale des Sciences d'Outre-Mer. Forthcoming, 2019.
Ramos, Afonso. "Angola 1961, o horror das imagens." In *O império da visão: A fotografia no contexto colonial português (1860–1960)*, edited by Filipa Lowndes Vicente, 397–432. Lisbon: Edições 70, 2014.
Reuver-Cohen, Caroline, and William Jerman, eds. *Angola: Secret Government Documents on Counter-Subversion*. Translated by Caroline Reuver-Cohen and William Jerman. Rome: IDOC, 1974.
Ribeiro, Nelson. *A emissora nacional nos primeiros anos do estado novo, 1933–1945*. Lisbon: Quimera, 2005.
———. "Broadcasting to the Portuguese Empire in Africa: Salazar's Singular Broadcasting Policy." *Critical Arts* 28, no. 6 (2014): 920–37.
Rodrigues, Eugénia. *A geração silenciada: A Liga Nacional Africana e a representação do branco em Angola na década de 30*. Porto: Edições Afrontamento, 2003.
Roediger, David R. *The Wages of Whiteness: Race and the Making of the American Working Class*. New York: Verso, 1991.
Rosas, Fernando. *História a história: África*. Lisbon: Tinta da China, 2018.
Rui, Manuel. *Crónica de um Mujimbo*. Lisbon: Cotovia, 1999.
———. *Quem me dera ser onda*. Lisbon: Cotovia, 1991.
Saíde, Alda Romão Saúte. "A *Voz da FRELIMO* and the Struggle for the Liberation of Mozambique, 1960s to 1970s." Paper circulated at the Wits Workshop on Liberation Radios in Southern Africa, University of the Witswatersrand, February 2017.
Sanches, Manuela Ribeiro, Fernando Clara, João Ferreira Duarte, and Leonor Pires Martins, eds. *Europe in Black and White: Immigration, Race, and Identity in the "Old Continent."* Chicago: University of Chicago Press, 2010.
Santos, Boaventura de Sousa. "Between Prospero and Caliban: Colonialism, Postcolonialism and Inter-Identity." *Luso-Brazilian Review* 39, no. 2 (2002): 9–43.

Scales, Rebecca P. "*Métissage* on the Airwaves: Toward a Cultural History of Broadcasting in French Colonial Algeria, 1930–1936." *Media History* 19, no. 3 (2013): 305–21.

———. "Subversive Sound: Transnational Radio, Arabic Recordings, and the Dangers of Listening in French Colonial Algeria, 1934–1939." *Comparative Studies in Society and History* 52, no. 2 (2010): 384–417.

Schubert, Jon. *Working the System: A Political Ethnography of the New Angola*. Ithaca, NY: Cornell University Press, 2017.

Schulz, Dorothea E. "Equivocal Resonances: Islamic Revival and Female Radio 'Preachers' in Urban Mali." In Gunner, Ligaga, and Moyo, *Radio in Africa*, 63–80.

Scott, James C. *Domination and the Arts of Resistance: Hidden Transcripts*. New Haven, CT: Yale University Press, 1990.

———. *Seeing Like a State: How Certain Schemes to Improve the Human Condition Have Failed*. New Haven, CT: Yale University Press, 1998.

Seekings, Jeremy. "'Not a Single White Person Should Be Allowed to Go Under': *Swartgevaar* and the Origins of South Africa's Welfare State, 1924–1929." *Journal of African History* 48, no. 3 (2007): 375–94.

Shadle, Brett L. *The Souls of White Folk: White Settlers in Kenya, 1900–1920s*. Manchester: Manchester University Press, 2015.

Shubin, Vladimir G. *The Hot "Cold War": The USSR in Southern Africa*. London: Pluto Press, 2008.

Spitulnik, Debra. "Documenting Radio Culture as Lived Experience: Reception Studies and the Mobile Machine in Zambia." In Fardon and Furniss, *African Broadcast Cultures*, 152–55.

———. *Media, Infrastructure, and Urban Culture in Nigeria*. Durham, NC: Duke University Press, 2008.

———. "Mediated Modernities: Encounters with the Electronic in Zambia." *Visual Anthropology Review* 14, no. 2 (1998–1999): 63–84.

———. "Mobile Machines and Fluid Audiences: Rethinking Reception through Zambian Radio Culture." In *Media Worlds: Anthropology on New Terrain*, edited by Faye D. Ginsburg, Lila Abu-Lughod, and Brian Larkin, 227–54. Berkeley: University of California Press, 2002.

———. "Personal News and the Price of Public Service: An Ethnographic Window into the Dynamics of Production and Reception in Zambian State Radio." In *The Anthropology of News and Journalism: Global Perspectives*, edited by S. Elizabeth Bird, 182–98. Bloomington: Indiana University Press, 2009.

Sterne, Jonathan. *The Audible Past: Cultural Origins of Sound Reproduction*. Durham, NC: Duke University Press, 2003.

Stockwell, John. *In Search of Enemies: A CIA Story*. New York: Norton, 1978.

Stoever, Jennifer Lynn. *The Sonic Color Line: Race and the Cultural Politics of Listening*. New York: New York University Press, 2016.

Stora, Benjamin. "The 'Southern' World of the *Pieds Noirs*: References to and Representations of Europeans in Colonial Algeria." In Elkins and Pedersen, *Settler Colonialism in the Twentieth Century*, 225–42.

Stremlau, Nicole, Emanuele Fantini, and Iginio Gagliardone. "Patronage, Politics and Performance: Radio Call-In Programmes and the Myth of Accountability." *Third World Quarterly* 36, no. 8 (2015): 1510–26.

Tacchi, Jo. "Radio and Affective Rhythm in the Everyday." *Radio Journal* 7, no. 2 (2009): 171–83.

Tali, Jean-Michel Mabeko. *Guerrilhas e lutas sociais: O MPLA perante si próprio (1960–1977)*. Lisbon: Mercado de Letras Editores, 2018.

———. "Jeunesse en armes: Naissance et mort d'un rêve juvénile de démocratie populaire en Angola en 1974–1977." In *L'Afrique des générations: Entre tensiones et négociations*, edited by Muriel Gomez-Perez and Marie Nathalie LeBlanc, 301–58. Paris: Karthala, 2012.

———. *O MPLA perante si próprio*. Vol. 2 of *Dissidências e o poder de estado (1962–1977)*. Luanda: Nzila Editora, 2001.

Tavares, Gonçalo M. *Breves notas sobre música*. Lisbon: Relógio D'Agua, 2015.

Thomaz, Omar Ribeiro. "Review of *Passagens para África*, by Cláudia Castelo." *Análise Social* 186 (January 2008): 183–90.

Thornton, John K. *A Cultural History of the Atlantic World, 1250–1820*. New York: Cambridge University Press, 2012.

Tiscar, Maria José. *A PIDE no xadrez africano: Angola, Zaire, Guiné, Moçambique; Conversas com o Inspetor Fragoso Allas*. Lisbon: Edições Colibri, 2017.

Tomás, António. "Becoming Famous: Kuduro, Politics and the Performance of Social Visibility." *Critical Interventions* 8, no. 2 (2014): 261–75.

———. "The Making and Unmaking of a Modernist City." Chapter 1 of unpublished manuscript. Cited with author's permission.

———. "Mutuality from Above: Urban Crisis, the State and the Work of *Comissões de Moradores* in Luanda." *Anthropology Southern Africa* 37, nos. 3–4 (2014): 175–86.

———. *O fazedor de utopias: Uma biografia de Amílcar Cabral*. Lisbon: Tinta da China, 2007.

Tomaselli, Keyan G., and Ruth Tomaselli. *Media Reflections of Ideology*. Durban: University of Natal, 1985.

Toogood, Alexander F. "Portuguese Dependencies." In *Broadcasting in Africa: A Continental Survey of Radio and Television*, edited by Sydney W. Head, 157–65. Philadelphia, PA: Temple University Press, 1974.

Torres, Silvia, ed. *O jornalismo Português e a guerra colonial*. Lisbon: Guerra and Paz, 2016.

Tostões, Ana, and Ana Maria Braga. "Unidade de vizinhança prenda Luanda à luz da Carta de Atenas." In *Arquitectura em África Angola e Moçambique*, edited by Ana Tostões, 164–87. Casal de Cambra: Caleidoscópio, 2008.

Travis, Trysh. "Books in the Cold War: Beyond 'Culture' and 'Information.'" In Auerbach and Castronovo, *Oxford Handbook of Propaganda Studies*, 180–200.

Trewhela, Paul. *Inside Quatro: Uncovering the Exile History of the ANC and SWAPO*. Auckland Park: Jacana, 2009.

Tudesq, André-Jean. *L'Afrique parle, L'Afrique écoute: Les radios en Afrique subsaharienne*. Paris: Karthala, 2002.

United Nations Center against Apartheid. International Commission of Inquiry. *Acts of Aggression Perpetrated by South Africa against the People's Republic of Angola, June 1979–July 1980*. New York: United Nations, 1981.

Vera Cruz, Elizabeth Ceita. *O estatuto do indigenato—Angola: A legalização da discriminação na colonização portuguesa*. Lisbon: Novo Imbondeiro, 2005.

Vidal, Nuno. "The Angolan Regime and the Move to Multiparty Politics." In Chabal and Vidal, *Angola*, 124–74.

———. “Social Neglect and the Emergence of Civil Society in Angola.” In Chabal and Vidal, *Angola*, 200–236.
Vidali, Debra Spitulnik. *See* Spitulnik, Debra.
Vieira, José Luandino. *A cidade e a infância: Estórias*. 3rd ed. Luanda: União de Escritores Angolanos, 1977.
Vigil, José Ignacio López. *Rebel Radio: The Story of El Salvador's Radio Venceremos*. Abridged and translated by Mark Fried. Evanston, IL: Curbstone Books, 1995.
Villegas, Harry. *See* Pombo.
Von Eschen, Penny M. *Satchmo Blows Up the World: Jazz Ambassadors Play the Cold War*. Cambridge, MA: Harvard University Press, 2004.
Vunge, Adebayo. *Dos mass midia em Angola: Um contributo para a sua compreensão histórica*. Luanda: A. Vunge, 2006.
Wald, Priscilla. “The ‘Hidden Tyrant’: Propaganda, Brainwashing, and Psycho-Politics in the Cold War Period.” In Auerbach and Castronovo, *Oxford Handbook of Propaganda Studies*, 109–30.
Westad, Odd Arne. *The Global Cold War: Third World Interventions and the Making of Our Times*. New York: Cambridge University Press, 2005.
Wheeler, Douglas L., and Réné Pélissier. *Angola*. New York: Praeger, 1971.
White, Luise. *Speaking with Vampires: Rumor and History in Colonial Africa*. Berkeley: University of California Press, 2000.
———. *Unpopular Sovereignty: Rhodesian Independence and African Decolonization*. Chicago: University of Chicago Press, 2015.
White, Luise, Stephan F. Miescher, and David William Cohen, eds. *African Words, African Voices: Critical Practices in Oral History*. Bloomington: Indiana University Press, 2001.
Willems, Wendy. “Participation— In What? Radio, Convergence and the Corporate Logic of Audience Input through New Media in Zambia.” *Telematics and Informatics* 30, no. 3 (2013): 223–31.
———. “Provincializing Hegemonic Histories of Media and Communication Studies: Toward a Genealogy of Epistemic Resistance in Africa.” *Communication Theory* 24, no. 4 (2014): 415–34.
Williams, Christian A. “Living in Exile: Daily Life and International Relations at SWAPO's Kongwa Camp.” *Kronos* 37, no. 1 (2011): 60–86.
Windrich, Elaine. *The Cold War Guerrilla: Jonas Savimbi, the U.S. Media, and the Angolan War*. New York: Greenwood Press, 1992.
———. “The Laboratory of Hate: The Role of Clandestine Radio in the Angolan War.” *International Journal of Cultural Studies* 3, no. 2 (2000): 206–18.
Wolfers, Michael, and Jane Bergerol. *Angola in the Frontline*. London: Zed Books, 1983.
Young, Crawford. *The African Colonial State in Comparative Perspective*. New Haven, CT: Yale University Press, 1994.
———. *Politics in the Congo: Decolonization and Independence*. Princeton, NJ: Princeton University Press, 1965.

INTERVIEWS

Alves, Fernando: 23 June 2015, Lisbon
Amaro, Rui. conversation, 24 May 2011, Luanda
Andrade, José "Zán." 29 June 2011, Lisbon
Assis, Albina. 17 January 2002, Luanda
Bandeira, Leston. 17 June 2011, Lisbon
Bento, Lourenço. 28 January 2011, Luanda
Borges, David. 15 May 2008, Lisbon
Cardoso, Oscar. 25 January 2016, Ericeira
Carreira, Ilda. 4 June 2008, Luanda
Carvalho, Fernão Lopes Simões de. 11 June 2013, Queijas
Carvalho, Rui Sergio de. 13 July 2011, Lisbon
Chaves, Sara. 17 June 2013, Barreiro
Costa, Emanuel "Manino." 25 May 2011, Luanda
Danda, Raúl Manuel. 10 February 2011, Luanda
Faria, Abelino "Manuel." 19 March 2002, Luanda
Ferramenta, Armando. 8 February 2011, Luanda
Ferreira, Carlos Monteiro "Cassé." 16 March 2011, Luanda
Fonseca, António. 23 April 2011, Luanda
Jaime, Alberto. 4 December 2001, Luanda
Jamba, Miraldina. 18 February 2011, Luanda
Jorge, Paulo. 2 August 2005, Luanda
Lara, Bruno. 17 April 2011, Luanda
Lomboleni, Filipe. 30 May 2011, Luanda
Malaquias, Bela. 6 June 2008, Luanda
Maria, Adolfo. 22 June 2005, Lisbon
Mogas, Guilherme. 11 and 24 May 2011, Luanda
Oliveira, José. 14 December 2015, Luanda
Paim, Eduardo. 9 May 2011, Luanda
Patrício, José. 13 June 2012, Luanda
Pereira, Elisabete. 11 April 2011, Luanda
Pirio, Greg. 18 August 2015, via Skype
Proença, Caro. 17 June 2013, Barreiro
Queiroz, Artur. 23 May 2011, Luanda
Roberto, Holden. 9 August 2005, Luanda
Sebastião, Matemona. 27 February 2002, Luanda
Silva, Ladislau (Lau). 16 February 2011, Luanda
Silva, Reginaldo. 24 January 2011, Luanda; and conversation 19 July 2018, Luanda
Silva, Sebastião. 20 April 2011, Luanda
Simons, Paula. 12 May 2011, Luanda
Traça, Mbeto. 9 May 2011, Luanda

Index

Page numbers in italics refer to illustrations.